EVANGELICALISM AND CONFLICT
IN NORTHERN IRELAND

Gladys Ganiel

CONTEMPORARY ANTHROPOLOGY OF RELIGION

*A series published with the Society for the
Anthropology of Religion*

Robert Hefner, Series Editor
Boston University
Published by Palgrave Macmillan

Evangelicalism and Conflict in Northern Ireland

Gladys Ganiel

First published in 2008 by
PALGRAVE MACMILLAN™
175 Fifth Avenue, New York, N.Y. 10010 and
Houndmills, Basingstoke, Hampshire, England RG21 6XS
Companies and representatives throughout the world.

PALGRAVE MACMILLAN is the global academic imprint of the Palgrave Macmillan division of St. Martin's Press, LLC and of Palgrave Macmillan Ltd. Macmillan® is a registered trademark in the United States, United Kingdom and other countries. Palgrave is a registered trademark in the European Union and other countries.

ISBN-13: 978–0–230–60539–8
ISBN-10: 0–230–60539–7

Library of Congress Cataloging-in-Publication Data

Ganiel, Gladys.
    Evangelicalism and conflict in Northern Ireland : by Gladys Ganiel.
        p. cm.—(Contemporary anthropology of religion)
    Includes bibliographical references.
    ISBN 0–230–60539–7
        1. Evangelicalism—Northern Ireland. 2. Northern Ireland—Church history. 3. Christianity and politics—Northern Ireland. I. Title.

BR1642.N73G36 2008
274.16′082—dc22   2007041198

A catalogue record for this book is available from the British Library.

Design by Newgen Imaging Systems (P) Ltd., Chennai, India.

First edition: June 2008

10 9 8 7 6 5 4 3 2 1

Printed in the United States of America.

For Marios and Eleanor Moussoulides

Contents

List of Tables

Acknowledgments

Special thanks are due to colleagues and students at the Irish School of Ecumenics, Trinity College Dublin, for their encouragement as I completed this project. Caroline Clarke, Amber Rankin, Jayme Reaves, and Emily Winsauer provided valuable administrative support. Robert Hefner, the editor of this series, was unfailingly helpful.

The research for this book originally formed the basis of a Ph.D. in the Department of Politics at University College, Dublin. I am indebted to my supervisor Jennifer Todd, whose insight, patience, and encouragement aided this work immensely. The Royal Irish Academy's Third Sector Research Programme provided the funding that made this work possible. The participants in the research were generous with their time and with practical assistance. The support of Brian O'Neill, Eleanor and Marios Moussoulides, and Ethel White removed considerable anxiety.

I also wish to acknowledge the insight and support of Claire Mitchell, Jean Brennan, John Brewer, John Coakley, Paul Dixon, Christopher Farrington, Claire Finn, Tom Garvin, Linda Hogan, Michael Kennedy, Laura Mahoney, Vincent O'Sullivan, Susanna Pearce, Bert Preiss, David Tombs, Max Bergman, and Veronique Mottier at the Essex Summer School, and my parents Carl and Jennie Ganiel. The Institute for the Study of American Evangelicals at Wheaton College, Illinois, provided a hospitable research home for six weeks in 2004; here I benefited from the assistance of Mark Amstutz, Leah Anderson, Amy Black, Larry Eskridge, Melissa Franklin-Harkrider, and Mark Noll. My Wheaton "family"—the Rowes: Maggie, Mike, Adam, Amber, and Jordan—were hospitable beyond the call of duty. The Walter Byers Award from the National Collegiate Athletic Association provided the funding for the pilot study that led to this research. Finally, a great many people at

Providence College could be held accountable for setting me on my journey to Ireland, but I must especially thank Ray Treacy and Andy Ronan, my cross-country coaches, Joseph Cammarano, my undergraduate thesis supervisor, and everyone in the Political Science Department who encouraged me to ask why.

List of Abbreviations

CAN	Christian Action Network
CARE	Christian Action Research and Education
CCCCW	Churches Central Committee for Community Work
CCCI	Centre for Contemporary Christianity in Ireland
CCF	Cooperative Commonwealth Federation
CCRU	Central Community Relations Unit
COI	Church of Ireland
CRC	Community Relations Council
DHSS	Department of Health and Social Services
DSD	Department of Social Development
DUP	Democratic Unionist Party
EA	Evangelical Alliance
EAPE	Evangelical Association for the Promotion of Education
ECONI	Evangelical Contribution on Northern Ireland
EFC	Evangelical Fellowship of Canada
EIPS	European Institute for Protestant Studies
EMU	Education for Mutual Understanding
EPS	Evangelical Protestant Society
ESA	Evangelicals for Social Action
IOO	Independent Orange Order
IRA	Irish Republican Army
ISE	Irish School of Ecumenics
IVCF	Intervarsity Christian Fellowship
NAE	National Association of Evangelicals
NDP	New Democratic Party
NICVA	Northern Ireland Council for Voluntary Action
NIHRC	Northern Ireland Human Rights Commission
NUP	National Union of Protestants
OFMDFM	Office of the First Minister and Deputy First Minister
OPC	Orthodox Presbyterian Church

PCI	Presbyterian Church in Ireland
PCRO	Peace and Conflict Resolution Organisation
PUP	Progressive Unionist Party
RUC	Royal Ulster Constabulary
SDLP	Social Democratic and Labour Party
SPRING	Society for the Promotion of Reformation in Government
SU	Scripture Union
UDA	Ulster Defence Association
UUP	Ulster Unionist Party
UWC	Ulster Workers Council
WEF	World Evangelical Fellowship
YWAM	Youth With A Mission

List of Previous Publications

Dr Gladys Ganiel
Irish School of Ecumenics, Trinity College Dublin

"Religion in Northern Ireland: Rethinking Fundamentalism and the Possibilities for Conflict Transformation," (2008), with Paul Dixon, *Journal of Peace Research*, 45(3).

"Preaching to the Choir? An Analysis of DUP Discourses about the Northern Ireland Peace Process," (2007), *Irish Political Studies*, 22(3): 303–320

"Religion and Transformation in South Africa?: Institutional and Discursive Change in a Charismatic Congregation," (2007), *Transformation: Critical Perspectives on Southern Africa*, 63: 1–22.

"Emerging from the Evangelical Subculture in Northern Ireland: A Case Study of the Zero28 and ikon Community," (2006), *International Journal for the Study of the Christian Church*, 6(1): 38–48

"Race, Religion and Identity in South Africa: A Case Study of a Charismatic Congregation," (2006), *Nationalism and Ethnic Politics*, 12(304): 555–576

"Turning the Categories Inside-Out: Complex Identifications and Multiple Interactions in Religious Ethnography," (2006), with Claire Mitchell, *Sociology of Religion*, 67(1): 3–21

"Ulster Says Maybe: The Restructuring of Evangelical Politics in Northern Ireland," (2006), *Irish Political Studies*, 21(2): 137–155

"Scandal and Political Candidate Image," (December 2000), with James Carlson and Mark S. Hyde, *Southeastern Political Review*, 28(4): 747–757

Book Chapters:

"A New Framework for Understanding Religion in Northern Irish Civil Society," in Christopher Farrington, ed., *Global Change, Civil Society and the Northern Ireland Peace Process: Implementing the Political Settlement*, Palgrave, 2008, under contract

"Religious Dissent and Reconciliation in Northern Ireland," in John O'Grady and Peter Scherle, eds., *Ecumenics from the Rim: Explorations in Honour of John D'Arcy May*, Berlin: LIT Verlag, 2007, 379–386.

Chapter 1

Introduction

Religion, Conflict, and Transition

Since September 11, the relationship between religion and conflict has become an urgent matter of consideration for citizens and policymakers. It has been all too easy to reduce this relationship to stereotypes about "fundamentalists" or "terrorists." This leads to simplistic conclusions: religion is dangerous and should be kept out of the public sphere; or the religious dimensions of conflict will disappear if society becomes more secular or modern (Ganiel and Dixon 2008).

This book challenges those conclusions, building on a growing body of international research that emphasizes the need to engage with—rather than dismiss—the religious dimensions of conflict (Little 2007; Tombs and Liechty 2006; Appleby 2000; Johnston 2003; Gopin 2000; Hefner 2000). It argues that the role of religion in conflict is always contextual and complex. Religion "matters" more in some conflicts than in others. And in conflicts where religion matters, it may matter more for some groups than for others. The challenge then becomes locating those for whom religion matters, understanding *how* religion matters for them, and exploring sociostructural factors and the specific religious beliefs that impact religious actors' ability to contribute to conflict transformation. This requires a multidimensional, multidisciplinary approach that utilizes conceptual and empirical insights from anthropology, history, political science, and sociology. With its focus on microlevel and incremental processes of change, anthropology provides an important and often overlooked perspective, without which an overall analysis of conflict and its transformation is incomplete. Anthropological data is especially useful for understanding the diversity within presumably monolithic and conflicting groups, and mapping social and political processes that may be taking place outside of political institutions or beneath the radar of the public sphere. Accordingly, this book uses an intensive, ethnographic study of evangelicalism in Northern Ireland to provide particular and concrete information about evangelicalism and its role in postconflict transition in Northern Ireland. It argues that anthropological approaches to religion can provide insights into other conflicts with religious dimensions and can be used to develop general, conceptual frameworks for understanding wider issues around religion, conflict, and transition.

In 1998, representatives of the Protestant/unionist and Catholic/nationalist communities in Northern Ireland agreed to a peace accord, referred to here as the Belfast Agreement.[1] The Belfast Agreement is the most comprehensive attempt thus far to achieve a settlement for the "Troubles" that escalated after the Catholic civil rights marches in the late 1960s. Since the Belfast Agreement, Northern Ireland has achieved some measure of stability and a marked reduction in violence. But significant problems remain. For instance, the power-sharing Assembly was suspended in 2002 and disagreements continue over policing and the failure of paramilitary groups to decommission, amongst other things. The Assembly was restored in May 2007, with what was once thought of as an unlikely combination of the Rev. Ian Paisley of the Democratic Unionist Party (DUP) and Martin McGuinness of Sinn Fein installed as first minister and deputy first minister. Paisley is a staunch evangelical who previously claimed that he would "never, never, never" share power with what the DUP calls "Sinn Fein/IRA." Relations between the DUP and Sinn Fein continue to be icy, and it remains to be seen whether these parties will be able to govern effectively.[2] Systemic aspects of the conflict have changed and loosened, but it is not yet clear whether Northern Ireland is truly in a postconflict or transitional phase.

The Troubles were not a religious war, but the religious dimension of the conflict often has been misunderstood (Mitchell 2006). Ruane and Todd's (2008, 1996) multidimensional theory of the conflict allows for an understanding of the place of religion within it. Their theory is rooted in a conception of conflict as a multidetermined, emergent property of a system of relationships on a number of levels. These levels—a set of overlapping differences; a structure of dominance, dependence and inequality; and a tendency toward communal polarization—interlock and mutually reinforce each other. The differences include religion, ethnicity, colonial status, culture, and national allegiance, each embodied in habitus. Religion is one dimension of difference that has contributed, in varying degrees at different points in time, toward the formation of polarized communities. Fulton (1991) also has focused on how religion, combined and overlapping with culture, economics, and politics, has been bound up with the power structures of each community, underwriting and legitimating communal conflict.

In Northern Ireland, evangelicalism has a historical prominence and a continued social significance in the unionist community that is not matched by Catholicism in the nationalist community. Religious dimensions to the community division and conflict have been more important for Protestant evangelicals (and by extension, the wider Protestant community) than for Catholics.[3] Addressing the religious dimension of the Northern Ireland conflict requires a comprehensive understanding of evangelicalism's historical and contemporary roles.

Evangelicalism has its origins in eighteenth-century revival movements in English Anglicanism and continental pietism. It was an international phenomenon, spreading quickly to North America through the missionary efforts of itinerant preachers such as John Wesley and George Whitefield. Evangelical history in Ireland is generally dated from 1747, when Wesley made his first visit to the island. Evangelicalism then, as now, was a diverse movement, encompassing Christians in a variety of denominations. Bebbington's (1989) fourfold definition of evangelicalism rings true for those original revivalists and those who would be considered evangelicals today. The characteristics that he identifies are that evangelicals believe that one must be converted or "born again"; that the Bible is the inspired word of God; that Christ's death on the cross was a historical event necessary for salvation; and that Christians must express their faith through social action/evangelism.

Although Bebbington's definition is widely cited and employed in the study of evangelicalism, evangelicalism is a contested concept and it is often conflated with fundamentalism. Indeed, "fundamentalism" is a term that some "evangelicals" (in Bebbington's sense) would use to describe themselves. Fundamentalism has its origins in an early twentieth-century movement within American evangelical Protestantism. The term is derived from *The Fundamentals*, a series of booklets published between 1910 and 1915. The booklets were designed to challenge "liberal" theology and biblical criticism (Marsden 1980). However, fundamentalism has come to have negative connotations and the term has been applied (often inaccurately) to movements across different religious traditions, especially Islam (Almond, Appleby, and Sivan 2003; Munson 2005, 2003).[4] This has meant that self-styled or self-defined evangelicals or fundamentalists have often taken great pains to distinguish themselves from one another.

This process of drawing distinctions between fundamentalists and evangelicals, or between different types of evangelicalism, takes place both in the academic literature and amongst people at the grass roots. It is a distinction that I have made previously (Ganiel 2002). When debates become particularly bitter—as they have been in Northern Ireland—rivals have accused one another of not being "really" evangelical. However, different types of evangelicalism share enough resemblance to be part of the same "family" (Thomson 1998, 1995b), and the historical origins of fundamentalism as a movement within evangelicalism cannot be denied. As such, Bebbington's four characteristics of evangelicalism are reference points in terms of which different types of evangelicals and fundamentalists define themselves, even if they hold different beliefs about what those characteristics mean to them.

The distinctions and struggles between different strands of evangelicalism and fundamentalism have on occasion proven to be socially and politically significant. In this research, I have chosen to distinguish between different types of evangelicalism, rather than between evangelicalism and fundamentalism, for two reasons. First, evangelicalism was an important movement in Northern Ireland before the emergence of fundamentalism from the American context. Second, evangelicals in Northern Ireland who could be classified as fundamentalists often call themselves evangelicals. Although evangelicalism has appeared quite unified during times of stress in Northern Ireland, there is significant diversity within it.

Exploring Evangelicalism
in Northern Ireland

An important part of understanding religion, conflict, and transformation in Northern Ireland is simply mapping what evangelicalism is like on the ground. Anthropological or ethnographic approaches are especially useful here for providing a comprehensive perspective on contemporary evangelicalism and how it is changing. The diversity within evangelicalism began to be documented and explored during the Troubles, and the distinctions within it have been defined in different ways by different scholars (Jordan 2001; Mitchel 2003; see also Brewer and Higgins 1998). This previous work and my own preliminary research led me to begin by conceptualizing what I have called traditional evangelicalism and mediating evangelicalism as empirical (or practical) categories for further investigation (Brubaker 2002). The fieldwork on which these categories are based was carried out from 2002 to 2005 and involved 61 in-depth, semistructured interviews with 57 individuals in four congregations and seven organizations.[5] I lived in Dublin at the time but traveled frequently to Northern Ireland for extended research visits. The scope of my study demanded a regional rather than single-site approach, that is, a single city or village (Lukens-Bull 2005:19). I currently live in Belfast, Northern Ireland, and continue to participate in and observe evangelical activities. I lived both in Belfast and Dublin whilst writing this book.

I based the empirical types on the religious beliefs that are prominent within Northern Irish evangelicalism and that are most important in divided societies: beliefs about the proper relationship between church and state, about religious or cultural pluralism, and about violence and peace. Although the types are informed by wider theological traditions (Calvinism and Anabaptism), the specific constructs are not intended as immediately generalizable. For example, views on violence or pluralism resonate differently in societies where violence is external rather than internal, or where pluralism is between different, rather than oppositional, groups. The traditional identity included Calvinist-informed beliefs about a covenantal relationship between the church and state, a privileged place for right religion and a selective use of violence as a last resort. The mediating identity included Anabaptist-informed beliefs about a strict separation between church and state, enthusiasm about religious and cultural pluralism, and nonviolence. The term mediating is borrowed from Noll's characterization of Canadian evangelicals as a group that have "moderated extremes" in Canadian politics and that interact with government and nonevangelical groups with an "accommodating spirit" (2001a:253).

I aimed to explore how people constructed their identities, made distinctions between themselves and others, and experienced change (both within and between categories). I conceived of the categories as open, fluid, and existing in the real world rather than as abstract "ideal" types. This allowed me to move beyond oversimple and stereotypical conceptions of evangelicalism. And by listening to the stories of the people who actually inhabit the categories, I aimed to deepen my understanding of the cognitive, normative, and practical content of these categories as understood by the participants. This provided a starting point for investigation, which I aimed to further specify and define as I gathered data. This eventually allowed me to construct distinguishable "empirical type" evangelical identities. In contrast to abstract and often restrictive ideal types, empirical types have the advantage of existing "as real types used by real people."[6] They are "systems of real belief and action...[with] an identifiable structure and form" (Brewer and Higgins 1998:132).

As my research progressed, I identified two additional categories. First, some evangelicals did not identify strongly either with traditional or mediating evangelicalism; rather, they indicated a desire to withdraw from society and politics. They were representative of a pietist category that has been well-documented both within Northern Irish and international evangelicalism (Jordan 2001; Mitchell 2003; Smith 1998; Reimer 2003). Second, I encountered people who identified as postevangelicals. Previous research did not lead me to expect to find people with postevangelical identities in Northern Ireland, but once I found them I could not ignore them. Postevangelicalism has been increasingly documented internationally (Gibbs and Bolger 2005; Rollins 2006; Ganiel 2006a; Tomlinson and Willard 2003; McLaren 2004). My task then became refining these broad categories into distinguishable empirical types.

The pietist identity implied a separation between church and state, and nonviolence. Pietist attitudes about pluralism were less easy to discern. Postevangelicalism included a separation between church and state, enthusiasm about religious and cultural pluralism, and nonviolence. It differed from mediating evangelicalism to the extent that it defined itself over and against traditional (and to some extent mediating) evangelicalism. In addition, postevangelical enthusiasm for pluralism extended beyond that of most mediating evangelicals. Postevangelicals may incorporate other Christian and religious traditions into worship and have a greater openness to new ways of interpreting the Bible. This typology allowed for people from a variety of denominations or with different party political preferences to be categorized within the same empirical type. The typology also highlighted often unnoticed similarities and key differences between categories of evangelicalism. Drawing out these distinctions allowed for a more nuanced understanding of the dynamics of change within and between categories. Table 1.1 summarizes the empirical-type identities.

These identities, constructed by each individual over a lifetime of socialization and experiences, provide the starting point for analysis. With this grounding, it is possible to begin explaining the changing role of evangelicalism in conflict and transformation. A complex web of variables are important for this, including the relationship between evangelicalism and sociopolitical power, evangelicalism's place in the structure of civil society (in particular, the extent that it is included in

Table 1.1 Empirical-Type Evangelical Identities

	Traditional	Mediating	Pietist	Postevangelical
Relationship between church and state	Covenantal	Separation	Withdrawal from society/ politics (Implied separation)	Separation
Pluralism	Privileged place for "right religion"	Advocates pluralism	Unclear	Advocates pluralism
Attitudes about violence and peace	Violence justified as a last resort	Nonviolence	Nonviolence	Nonviolence

civil society or not), the effectiveness of religious structures (such as congregations or organizations), and the content of religious beliefs. This book analyzes these aspects in detail, focusing on the mechanisms of identity change amongst evangelicals in congregations, as well as on the changes in the way activists in evangelical organizations "frame" or justify their sociopolitical projects. It does not argue that any one variable in the web is the most important for explaining change; rather, the factors are interrelated and their relative importance for explaining change varies within individuals, congregations, and organizations. Further, the research identifies examples of actual identity change and the reframing of sociopolitical projects and includes evidence that these changes are affecting the process of conflict transformation. For instance, changes amongst mediating and postevangelicals point to a loosening of oppositional identities and to enthusiasm for "antisectarianism" projects. Even traditional evangelicals—who are usually opposed to the Belfast Agreement—have changed the way they frame their sociopolitical projects. They are now arguing in terms of discrimination, inequality, and their right to a "place at the table" in civil society. This indicates at least a practical acceptance of some of the principles of the Belfast Agreement.

Comparative perspectives shed further light on the possible future directions for evangelicalism and the possibilities for conflict transformation. The research cannot *predict* the future role of evangelicalism in Northern Ireland, but by examining how similar processes of change have occurred in the USA and Canada, it establishes a socioscientific basis for more general hypotheses about the possibilities. The USA and Canada, like Northern Ireland, are societies in which evangelicalism once held a privileged position. As evangelicals negotiated the breakdown of their relationships with power in those contexts, they were both enabled and constrained by emerging sociopolitical structures. For instance, in the USA, evangelicals engaged in active interest-group activity and focused on influencing the Republican Party. Movement in these directions was aided by the USA's diffuse, federal political structures. In Northern Ireland, these are the tactics favored by traditional evangelicals. Local government reforms and the reinstitution of the Assembly would favor these strategies. In Canada, evangelicals were more likely to reject interest-group politics, opting to try and influence government ministers and various political parties in a discrete, behind-the-scenes manner. Movement in this direction was aided by Canada's centralized political structures. In Northern Ireland, these tactics have been used quite effectively by mediating evangelicals under the British government's Direct Rule structures. Their strategies would be facilitated by a centralized government structure that continues to emphasize aspects of the British government's approach to grassroots conflict resolution, such as a preference for cross-community activists and organizations.

Structure of the Book

Conflict transformation is conceived of as the process whereby a postconflict or transitional society undergoes radical changes that mitigate or eliminate causes of the conflict. This involves changing relationships between previously antagonistic groups and dismantling unequal social, political, and economic structures. A core argument of this book is that the role of religion in this process is often overlooked, thus limiting the potential for transformation. The overlooking of religion is related to a tendency of theorists and policymakers to accept the assumption, embedded in modernization theory, that religion is a private matter that should be kept out of the public sphere. But as Hefner (2000) and others have pointed out, modernization has been accompanied by religious revivals and it does us no good to ignore religious ideas and actors. This book provides general, conceptual frameworks for understanding how religion contributes to conflict and transformation and analyzes how such processes are being worked out in Northern Ireland. It argues for a multidimensional and multidisciplinary approach to understanding religion and conflict, highlighting how anthropology's distinctive focus on microlevel processes provides often-overlooked insights that are of utmost importance to scholars and policymakers.

Chapter 2 draws on theoretical approaches to civil society, conflict management, and conflict transformation, constructing a framework for understanding how civil society functions in conflict situations. An anthropological perspective on civil society informs the treatment of the wider social science debate, leading to a redefinition of civil society in terms of its functions rather than its institutions. It includes analyses of processes of transformation, such as identity change and the reframing of sociopolitical projects. It also locates religion within civil society, analyzing how various religious structures function. It concludes by analyzing the changing structure of Northern Irish civil society, including the strategies underlying the British government's approach to conflict management and the extent that religion is included in the public sphere.

Chapter 3 analyzes the role of evangelicalism in Northern Ireland over time, exploring how the breakdown of the relationship between evangelicalism and power has created the space for evangelicalism to contribute to conflict transformation. It also uses a comparative approach to shed light on how evangelicals have continued to negotiate the breakdown of their relationship with power since Direct Rule. By analyzing the factors that led to the emergence of new forms of evangelical activism in the USA and Canada, it establishes a socioscientific basis for understanding the present modes and potential future trajectories of evangelical activism in Northern Ireland.

Chapters 4 and 5 draw on fieldwork, exploring how evangelicals operate through well-developed networks of congregations and special-interest organizations. Chapter 4 is based on fieldwork carried out in an urban Free Presbyterian congregation, an urban Presbyterian congregation, a rural Presbyterian congregation, a rural Church of Ireland (Anglican) congregation, and a postevangelical "community." It argues that changes in evangelical identities are reinforcing and shaping wider trends within the Protestant community, including withdrawal from the public sphere, assimilation, or conversion to the new political order.

Chapter 5 is based on fieldwork with three traditional evangelical organizations (the Evangelical Protestant Society, the Caleb Foundation, and the Independent Orange Order) and four mediating and postevangelical organizations (Evangelical Contribution on Northern Ireland, Evangelical Alliance, ikon, and Zero28). It argues that traditional evangelicals have reframed their activism in terms that signify at least a partial change in traditional evangelical ideals. Rather than arguing for a privileged place for "right religion," they make their case for their inclusion in the public sphere on the grounds that they are facing marginalization and discrimination. They focus on "moral" issues such as abortion and homosexuality legislation rather than issues such as preventing a united Ireland. Their decision to reframe their sociopolitical projects in terms of moral issues confirms their acceptance of the current context. In contrast, mediating and postevangelicals are enthusiastic about the changes that have occurred. They claim that their work has contributed to changes in identity amongst evangelicals at the grass roots. They reframe their activism in terms that indicate a significant break with traditional evangelical ideals. They have abandoned covenantal Calvinist assumptions about the relationship between church and state, not because they are no longer realistic—but because they think they are wrong. They urge evangelicals to repent for the way that evangelicalism has been bound up with power in Northern Ireland, believing that this is necessary for forgiveness and reconciliation in the wider society.

Some general conclusions that follow from these considerations are analyzed throughout the book and developed more fully in the final chapter. First, conflict transformation does occur, and it is possible to analyze how processes of transformation take place. This research provides evidence that evangelicals are participating in these processes—from the mediating and postevangelicals who self-consciously work for conflict transformation to even some traditional evangelicals. This debunks the popularly-held notion that in conflicts with religious dimensions, the most deeply religious people are obstacles to peace. Rather, it argues that politics motivated by religious convictions are more important for moving beyond conflict than is commonly supposed. There is a reason for this: if identity change is to occur, it must appeal to something beyond existing binary or oppositional identities. The sort of overriding convictions involved in religion—especially if they are coupled with critiques of the way religion has been used in the past to reinforce conflict— provide this. This is not to say that evangelicalism is the only or even the prime vehicle for change in Northern Ireland. But as in so many conflicts with religious dimensions, evangelicals' contributions are vital for facilitating wider processes of transformation.

Chapter 2

Civil Society, Religion, and Conflict in Northern Ireland

This chapter develops conceptual tools for understanding religion and conflict in divided societies. This provides a template for understanding processes of change in Northern Ireland, and for drawing wider, general conclusions about other conflicts with religious dimensions. All too often, theoretical approaches to civil society have reduced it to "associational life" or assumed that it unproblematically produces good citizens or social capital (Hefner 2001, 1998; Hann 1996). However, an anthropological perspective on civil society is used to refine and reconceptualize the wider social science debate, leading to a redefinition of civil society in terms of its functions rather than its institutions. These functions—contributing to the process of socialization and the practice of nongovernmental politics—provide scope for understanding how civil society may be a site for both conflict and transformation.

The latter part of the chapter deals with how some of these concepts have been reflected in approaches to conflict management in Northern Ireland. This includes analysis of how cross-community peace and reconciliation organizations (PCROs) have attempted to use civil society as an arena for conflict transformation; and how the British government has attempted to use civil society for conflict management through institutions such as the Community Relations Council (CRC) and the Civic Forum. These approaches have had a partial impact on the transformation of the conflict, but their results have been limited. Crucially, the extent to which religion is included on equal footing in the public sphere is not clear. This chapter argues that the inclusion of religion is a vital aspect of a wider process of conflict transformation.

Defining Civil Society

Civil society is one of the "big ideas" of the early twenty-first century (Edwards 2004; Hefner 1998). It is usually considered a vital component of a healthy democracy, without which polities are democratic in name only. But there is surprisingly little agreement about what exactly it is. Indeed, it is easier to say what it is not. Civil society is not the state, supranational institutions, the market, or the private sphere of the home or family (although some analysts have included the market and the family within civil society). Most definitions conceive of it in terms of "associations," and/or the "public sphere." A potentially infinite number of associations ranging from sports clubs to churches to trade unions are considered part of civil society. However, there has been a tendency for civil society theorists to exclude those groups from civil society that do not conform to liberal norms of "civility." This has led to the inevitable questions of "who decides" and has severely limited the ability of civil society theories to serve as useful conceptual tools in conflictual or divided societies (Hefner 2001).

Anthropologists have been critical of the concept of civil society, pointing to its Western origins and the ethnocentric assumptions of many of its proponents (Hann 1996). For example, in his study of democracy and culture in China and Taiwan, Weller refuses to use the term civil society, preferring to analyze the prospects for "democratic civility" (1999:16). For him, Western theorists have ignored the importance of "communal kinds of ties" that do not fit with narrow definitions of "associations" or "organizations." These theorists also overstate the "dichotomy between state and society" (Weller 1999:16). Similarly, in her study of Islamist mobilization in Turkey, White (2002) develops the concept of "vernacular politics" in an attempt to move beyond the civil-society–versus-state frameworks imposed by Western theories. Putting a twist on the insights developed by Putnam in his theory of social capital, White emphasizes the importance of the links between local culture, interpersonal relations, and community networks, which are "connected through civic organizations to national party politics" (2002:27). Hann (1996:3) sums up what he calls the most important general points about the civil society debates in the social sciences:

1. that civil society debates hitherto have been too narrowly circumscribed by modern Western models of liberal-individualism, and
2. that the exploration of civil society requires that careful attention be paid to a range of informal interpersonal practices overlooked by other disciplines

Hann argues that social scientists should "be prepared to abandon this universal yardstick [of civil society], and to understand civil society to refer more loosely to the moral community, to the problems of accountability, trust and cooperation that all groups face" (1996:20). Anthropological approaches can thus both provide a way to avoid some of the conceptual pitfalls critiqued here, and open up new avenues for exploring social and political relationships at the grass roots. This can be accomplished by conceiving of civil society in terms of its functions rather than its institutions. Such an approach draws on the classical theories of civil society prominent in the Western social sciences, whilst at the same time taking on board the anthropological critiques of Hann, Weller, White, and others. For example, most of the functions of civil society can be subsumed under two broad areas: the process of socialization and the practice of nongovernmental politics. Anthropological perspectives on civil society offer especially useful insights when considering the process of socialization, whilst the classical theories of civil society have been more prominent in academic analyses of the practice of nongovernmental politics.

In situations of conflict or divided societies, however, it is important to move beyond the assumption that the functions of civil society always have positive consequences. This assumption has been dominant both in the West and in non-Western contexts, where civil society is often seen as the arena for democratization (Hefner 2001). For example, socialization is conceived of as a process in which associations, the family, and other forms of organized life produce "good citizens." Edwards classifies these kinds of conceptions as "civil society as the good society" (2004:37–53). Kaldor's (2003) *societas civilus* (civil as an adjective) and *bürgerliche Gesellschaft* (Hegelian and Marxian ideas that include all organized life between the state and the family) could also be included in these conceptions.[1] Putnam (2000, 1993) is one of the most heralded and influential proponents of the normative worth of the process of socialization. Putnam and his followers argue that participation in organizations builds citizenship skills and "social capital." Social capital refers to the bonds that develop amongst citizens through social interaction. These bonds help to produce a climate of trust in which community and economic development flourishes. Likewise, the practice of nongovernmental politics is viewed as an empowering and effective means for citizens to shape the society in which they live. Edwards' classifications of "civil society as associational life" (2004:18–36) and "civil society as the public sphere" (2004:54–71) encompass these concepts. In Kaldor's terms, these conceptions are "activist" or "neoliberal." They draw on a de Tocquevillian heritage that focuses on how citizens organize into associations to communicate their concerns to the government. Civil society then works as a check against government becoming unjust or oppressive, and as an arena in which innovative social and political ideas are developed. They also draw on a Habermasian heritage that emphasizes the importance of debate in the public sphere. When civil society groups compete amongst themselves, offering radically different points of view, they stimulate critical debate that results in decisions that promote the common good.

That is all very well and good—in theory. But a more coherent and realistic analysis of the functions of civil society includes both their positive and negative potential (Hefner 2001; Keane 1998, 1996, 1988a, 1988b; Cochrane 2004, 2001; Little 2004; Edwards 2004; Tempest 1997). As Hefner observes, "nary a word is said about how civic associations may be cross-cut by deep ethnic, religious or ideological divides" (2001:9). In such cases, participation in organizations may produce conflict and distrust rather than trust and social capital. People may be socialized into segregated, competing communities that "bond" in a way that includes negative or oppositional conceptions of other communities (Varshney 2003, 2001). As far as the practice of nongovernmental politics is concerned, the competition of opposing groups may produce conflict rather than policies geared toward the "common good." Groups may develop relationships with government in which they become dependent on it for funding or support. Then, civil society may lose its ability to work as a check against the government and may contribute to unjust policies. These dangers are acute in divided societies, where there is little agreement about what constitutes a common good and in which there may be great disparities in opposing groups' relationships with the government. For civil society to play what Hefner calls a "democracy enhancing role," or for it to contribute to a peaceful postconflict transition, "the discourse and practice of people in public associations must be politically and culturally *civil....* Only when this *cultural quality* of 'democratic civility' is added to the *structural reality* of civic association can we say that 'civil society' has begun to do the job of strengthening democracy" (Hefner 2001:10).

The fact that civil society has both positive and negative functions means that it is an important variable that affects the course of conflicts and the success of political settlements. Conflict management must confront civil society head on, rather than pushing it to the periphery by focusing on elite deals and pretending that what goes on in civil society does not really matter. This means addressing the questions of whether and how government, or government and civil society actors working in tandem, can restructure civil society in such a way that it becomes a more hospitable environment for conflict transformation. Strategies for managing conflict must go beyond the more simplistic accounts of civil society enthusiasts, who at times imply that what goes on in civil society is more important than politics and that government is a largely malign influence that only hinders the grass roots from achieving peace on their own. Hefner's (1998) five lessons on "democratic civility" can be adapted to serve as a useful guide for exploring the potential of civil society in postconflict transitions:

1. Civic values of equality, participation, and tolerance must be embedded in a "broadly based political culture" and "scaled up" into the values of the state (Hefner 1998:15).
2. A self-limiting "civil state" must guarantee citizen participation.
3. Indigenous grassroots values and practices (including religious ones) must be cultivated for their potential to contribute to civic life and discourse.
4. There must be a "sphere of uncoerced association, speech and exchange in which different ideas of the good can be debated and tried" (Hefner 1998:27).
5. "Democratic civility depends upon cultural and institutional embedding, the precise structure of which varies from society to society"; a universal model of civil society should not be imposed without taking into account the particulars of each context (Hefner 1998:35).

Hefner's five lessons draw on anthropological, historical, political, and sociological approaches to civil society. Below, some conceptual tools from these disciplines are analyzed in light of their usefulness for understanding civil society in divided societies.

Conceptual Tools

Strategies for Managing Conflict

Civil society is a potential site for conflict because people can be socialized into competing, oppositional blocs. It also can foster inequalities amongst groups in their access to government patronage and the public sphere. These difficulties are usually overlooked in Western liberal conceptions of civil society, which assume that in a society based on procedural rules, "the problem of setting up a state can be solved even by a nation of devils" (Kant, quoted in Sandel

1996:322). However, the liberal paradigm has been subjected to criticism as the difficulties of keeping the devils in check become increasingly clear, especially in multicultural states (Kymlicka 1995; Little 2004; Keane 1998). Theorists have attempted to deal with this by offering frameworks about how to manage conflict within civil society. These are based upon addressing the aspects of conflict those theorists deem most important. For example, multiculturalists develop their frameworks based on the premise that clashes amongst opposing identities are most important. Feminists emphasize the importance of inequalities and devise solutions to eradicate them. These theories are usually presented as competing frameworks. This leads to a failure to recognize that the relative importance of each aspect of conflict will vary from context to context. Accordingly, rather than trying to adjudicate between these frameworks, a better strategy is to view them as incremental, identifying aspects of conflict that can have more or less salience in different contexts. The more important task, then, is tailoring conflict-management strategies to fit particular contexts.

For example, in contexts where people are socialized into oppositional blocs, there may not be enough "shared values" to promote social unity, or the divisions are so deep that any "shared values" cannot overcome them (Kymlicka 1995:187; Parekh 2000; see also Keane 1996). This may make reaching consensus impossible, and some of the values of competing groups just may never be reconciled (Mouffe 2000; Little 2004). Multicultural theorists have attempted to deal with this by focusing on the politics of identity or difference. For Kymlicka this involves citizens learning to possess both their primary identity *and* a "shared identity" (1995:188). He argues that the various ethnic, national, or cultural identities must be "nurtured rather than subordinated" before a shared identity can develop (Kymlicka 1995:189). He recommends an educational program that affirms existing multicultural identities and at the same time constructs a shared identity. This program would emphasize interethnic, cultural, and community contact as a means of promoting the value of diversity (Varshney 2003, 2001). Kymlicka also recommends institutional safeguards such as proportional representation, a self-government option, and affirmative action policies. Parekh (2000) argues along similar lines. He calls for a focused, educational program about multiculturalism; the public recognition of identities in social and political institutions; the establishment of collective rights for all groups; an agreed system of justice and policing; a minimal constitution safeguarding basic rights; and the nurturing of a "common culture." Parekh argues that structures of authority and the "common culture" should be viewed as contingent and open to change, rather than as essential or absolute. This gives people more flexibility in dealing with the tensions between particular and shared identities.

Radical democrats agree with multiculturalists that competing identities and viewpoints must be allowed to exist in the public sphere. Their strategies for dealing with conflict and division differ in that they want minimal institutional safeguards, their strategies are less concrete, and they deliberately avoid questions of institutional design. This is intentional, because it is meant to avoid focusing needlessly on theories that do not work in the "real world." This involves thinking of the political (in particular, political institutions or agreements) as always contingent, contested, and open for negotiation. The strategy, then, is changing mindsets so that people begin to conceive of it as a good thing for political institutions and agreements to be open to criticism and change.

In contexts where there is great inequality within civil society, or where some groups are excluded from the public sphere, other strategies may be necessary. Multiculturalists address these issues by arguing for the equal institutional recognition of cultures or nationalities. Feminists have countered that this does not go far enough, highlighting the exclusion of women and the working classes. Keane (1998), a self-identified "postfoundationalist,"[2] seeks to make civil society more inclusive by cultivating "public spheres of controversy." These are "spaces" where citizens (often the victims of violence) can monitor the violent tendencies of others. Publicizing violence is meant to help citizens remember past violence in order not to repeat it, to expose current violence, and to formulate new means for eliminating it (Keane 1998:156). He advocates focusing on the effects of violence on victims—through truth commissions, for example—to help counterbalance the "benefits" that perpetrators gain through publicity. More controversially, Parekh (2000) has argued that liberal society has attempted to exclude religion from the public sphere and that it should be included. The inclusion of all groups engaged in conflict—even contentious ones such as religious fundamentalists or paramilitaries—can be seen to address some of the problems and grievances that contributed to conflict in the first place.

Conflict Transformation, Socialization and Identity Change

In societies dominated by opposing identities, transformation may occur when the process of socialization and identity formation is disrupted, leading to changes in identity. Todd's (2005) typology provides a framework for understanding the mechanisms of this process. Drawing on Bourdieu, her approach takes into account wider social processes and resource distribution, as well as individual perceptions of choice. Todd recognizes that collective identity categories are not monolithic and that there are many different ways to construct them.[3] Taking these complexities into account, Todd identifies "three mechanisms of change in collective categories of identity" (2005:438):

- dissonances between the social order and the individual habitus
- dissonances within the individual habitus
- the moment of intentionality in identity formation

The dissonances between the social order and the habitus and within the habitus intersect with "moments" of intentionality in the ongoing process of identity construction (Todd 2005:434). Radical sociopolitical changes (such as those that have occurred in postconflict settings) may bring collective identity categories into dissonance or contradiction with the new order. This forces individuals to "re-sort the elements of their identity" (Todd 2005:439). This may lead to a moment of intentionality in which individuals decide to break with old aspects of their identity and to embrace new ones. Accordingly, Todd identifies six possible directions of identity change: reaffirmation, assimilation, conversion, adaptation, ritual appropriation, and privatization. Three of Todd's possible directions involve "transparency" between category and practice. Reaffirmation involves recognizing and affirming existing binary identities and resisting changes in their names. This may lead people to (violent) protest, or to marginalization. Assimilation involves reshuffling aspects of identity, rejecting some aspects of the old binary identities, and "placing other categories closer to the centre of identity" (Todd 2005:443). This may lead to an at least reluctant acceptance of the new order. Conversion means that people come to see their old identities as irrelevant in the face of structural changes, and to accept "the symbolic grammar embedded in the new order" (Todd 2005:441). This often signals an enthusiastic acceptance of the new order. Her other three directions are more ambiguous. Adaptation involves holding onto core aspects of identity, but in practice adapting to the new order so that contradictions arise between practice, meaning, and value. These contradictions may result in a tendency toward "crises" (Todd 2005:450). Ritual appropriation means accepting new practices

Table 2.1   Typology for Identity Change

No Change	Some Change	Transformative Change
May lead to marginalization or (violent) protest.	Focus on aspects of identity that are relevant to the new order, other aspects fade into the background. Leads to acceptance of the new order.	Leads to acceptance of the new order.
Core identity remains in contradiction to new order. Results in crises.	Conceptual contradictions and ambiguities may contribute to crises.	Withdrawal and de facto acceptance of new order. May be an interim phase where people decide to test new identities or to reaffirm old ones.

and redefining or assimilating them to fit with old narrative forms, rituals, and identities. This also may lead to some conceptual contradictions or ambiguities, contributing to crises. Finally, privatization involves people seeing their old identities as irrelevant in the face of change. But rather than accept those changes, they retreat into isolated, private worlds. This may signal a de facto acceptance of the new order; or it may be "an interim phase, a period when identity change too difficult or dangerous to accomplish publicly may be tested and controlled in a safe environment" (Todd 2005:448).

Todd's framework allows us to see how identities change, and how identity change leads people to reject, accept, or adjust to new social and political dispensations. To the extent identity change drives the process whereby people either enthusiastically or pragmatically adjust to the new order, it contributes to the transformation of conflict. Table 2.1 summarizes possible directions of identity change.

Conflict Transformation, Nongovernmental Politics and the Reframing of Sociopolitical Projects

Another way in which conflict at the level of civil society may be transformed is when activists redefine or "reframe" sociopolitical projects. Theoretical approaches to new social movements that focus on "framing processes" contribute to the understanding of how this takes place (McAdam, McCarthy, and Zald 1996; Melucci 1988a, 1988b, 1985; Eyerman and Jamison 1991; Touraine 1978). From this point of view, achieving concrete social or political goals may be important, but it is not as important as the way in which activists question old ways of thinking, justify their new ways of thinking, and disseminate these ideas in the wider society. Social movements (in particular, social movement organizations) are seen as carriers of these new discourses and as instrumental in stimulating change—both in the identities of movement participants and in the wider society. These processes may be observed even if there is not an obvious, full-blown social movement taking place. Social movements often pass through latent phases, when new strategies are being formed and new ideas are emerging. Focused research can identify the nongovernmental politics that organizations are engaging in and map the discourses that they are using to frame their activism.

This is especially important in postconflict settings, where radical sociopolitical changes may bring into question previous justifications for practicing nongovernmental politics. There may be dissonances between the discourses that were previously used to justify social activism and the acceptance of those discourses in the new public sphere. This may lead activists to criticize and abandon the old discourses, enabling them to participate fully in the new public sphere. Or, they may cling to the old discourses even in the face of public opposition, risking marginalization. Finally, they may retain elements of the old discourses while selectively adapting them to fit in the new public sphere. This would allow them to engage in the public sphere on select issues. These responses, and their possible consequences, can be summarized as shown in table 2.2.

This allows us to see how wider changes may prompt organizations to reframe their sociopolitical projects. This is more significant than whether or not they achieve concrete, measurable goals—especially if their redefinitions change discourses and perceptions in the wider public sphere. To the extent that reframing contributes to a process whereby people either enthusiastically or pragmatically participate in the public sphere, it contributes to the transformation of conflict.

Table 2.2 Typology for Reframing Sociopolitical Projects

No Change	Some Change	Transformative Change
Use the old discourses. Risk marginalization.	Retain the old discourses, while simultaneously adopting new discourses that are acceptable in the new public sphere. Risks presenting a "contradictory" stance but able to participate partially in the public sphere.	Criticize and abandon the old discourses, while simultaneously adopting new discourses that are acceptable in the new public sphere. Able to participate in the public sphere.

Limitations

The strategies outlined above are not foolproof, and there is no guarantee that processes of conflict transformation will occur. For instance, the strategy of constructing a shared identity or a common culture may be much more difficult than some multiculturalists imply. Traditional sources of a shared identity or a common culture (history, language, religion) are usually problematic in divided societies. Although Kymlicka cites Charles Taylor for arguing that building a society based on "deep diversity" might be a source of unity and "an object of pride," Kymlicka asks (perhaps more realistically) why this might not just be wearying instead! He says Taylor's project is unlikely to work "unless people value deep diversity itself, and want to live in a country with diverse forms of cultural and political membership" (Kymlicka 1995:191). It is not always clear that this is the case, or it may take a long time for people to conceive of society in these terms. Radical democrats such as Little (2004) point out that the theories of multiculturalists such as Kymlicka and Parekh have developed out of "real world problems," and that they can not be universally applied to contexts that are divided in a different way. For example, Northern Ireland, which is divided between "traditions" or "communities," would not fit Kymlicka's criteria of a "multinational society" or "the definition of culture that Parekh employs" (Little 2004:85–86).[4] In particular, Little is troubled by the institutional safeguards recommended by multiculturalists because they focus on the authority of the state to enforce them. In a society divided like Northern Ireland—where the nationalist community has not always accepted the authority of the state—this becomes problematic. It follows that Little argues for fewer institutional safeguards, especially if the state must be depended on to enforce them.

Although Little argues that this way of thinking about the political is realistic and useful, such a minimalist approach to institutional safeguards falls into the same trap as theorists of civil society who assume that competing groups will be "nice" to each other. Radical democrats do not convincingly demonstrate that a "different way of thinking" would be better at managing conflicts than the more concrete institutional safeguards offered by multiculturalists. Indeed, when Little criticizes Kymlicka and Parekh for tailoring their strategies to specific contexts, he is in fact criticizing them for doing what he claims radical democrats supposedly do best: taking a flexible approach to the "realities" that are present on the ground.

The question of what groups are included in or excluded from the public sphere is also problematic. Feminists and radical democrats have been most vociferous in arguing that the public sphere should be open to all points of view. For instance, Little argues that groups associated with paramilitarism should not be indefinitely excluded, and that the state and other civil society actors should be flexible enough to allow their participation at various stages (Little 2004:184–189). However, Little also asserts that religious groups *should* be excluded from the public sphere, claiming that the churches should not have more influence on public life in Northern Ireland. He says that, at most, non-Christian religious communities should be provided with a public platform. It is telling that Little, arguing from a radical democratic perspective that claims to welcome all stances (even the "uncivil" ones), thinks that religious viewpoints and identities are the only ones that have been so "divisive" that they should be cordoned off from the public sphere. He even concludes that it is "almost impossible" for religion to make a positive contribution in Northern Ireland (Little 2004:76). This shows a misunderstanding of the role of religion in societies where religion has been a dimension of conflict. Excluding religion from the public sphere essentializes religion, denies that religious identities can change, and underestimates the impact that religion can have in conflict transformation.

This is ironic, given that one of radical democrats' main criticisms of multiculturalists is that they have essentialized differences in identity and culture, making it more difficult for change to take place. It may be that the multiculturalists' strategies and safeguards for managing conflict mitigate its transformative potential, limiting the options that present themselves to people as they negotiate changes in their identities and contemplate new sociopolitical projects. Although the state may offer incentives to shape less oppositional identities and to build a shared identity, there is no guarantee that this will work. The radical democrats' emphasis on flexibility (and, indeed, the flexibility reflected in Kymlicka and Parekh's application of their theories to real-world contexts) may alleviate this tendency. The limitations of the strategies for managing conflict and the limitations of transformative processes must be finely balanced against each other, and this is notoriously difficult to do.

Religion in Civil Society

The role of religion in civil society is controversial. There are those who consider religion inconsequential or argue that its influence is waning with every passing day (Bruce 2001a, 1995; Berger 1967; Wilson 1979). This "secularization paradigm" was once dominant in the social sciences (Martin 1978). However, a growing body of work has focused on the persistence of religion; and, in particular, its continued role in the public sphere (Davie 1994; Casanova 1994; Smith 1998; Stark 1999; Herbert 2003; Dionne and Diiulio 2000). There are also those who argue that religion is a private affair and that including it in the public sphere only brings division and strife (Cohen 1998; Greenawalt 1996; Rawls 1993). They are challenged by others claiming that religion has a unique and positive role to play in public debate (Appleby 2000; S. Carter 1993; Neuhaus 1986; Niebuhr 2001 [1951]; Gopin 2000; Johnston 2003; Levinson 1992; Hessel 1993). In divided societies, religion's role is even more controversial.

As a part of civil society, religion has the potential to be an influence that is both a malign and benign. However, there are certain structural and ideational factors that impact how religion functions in any given context. These include the relationship between religion and sociopolitical power, the structure of civil society and religion's place in it, how different religious structures function, and the relationship between religious belief and sociopolitical action.

Religion and Power

The role of religion in any given context depends in large part upon its relationship with sociopolitical power. Jelen and Wilcox (2002), drawing on Leege (1993), divide the role of religion into two broad categories: the "priestly" and the "prophetic." In priestly religious politics, church and state "stand in a mutually supportive relationship to one another" (Jelen and Wilcox 2002:7). In other words, religion is bound up with sociopolitical power. In prophetic religious politics, "political and religious authority may assume opposed or independent roles" (Jelen and Wilcox 2002:7). In these cases, religion operates relatively independent of sociopolitical power.

Jelen and Wilcox frame their argument in terms of religion's relationship with the state. The state may use its power to support a particular religion, privileging it above others. This is the case when there are officially established religions, or when a particular religion is numerically dominant and receives unofficial support from the state. In these situations, religion is more likely to play a priestly role, providing moral or divine justification for the state's authority. Religion functions as "social cement" or as "sacred canopy," binding people together and providing meaning for group or national identity (Berger

1967; Durkheim 1968; Demerath 2001).[5] As such, a close relationship between religion and the state may promote solidarity, integration, and civic values. On the other hand, a close relationship between religion and the state may mean that religious actors fail to notice if the state is abusing its political power. In such cases, religion becomes the "opium of the people," blinding the oppressed to the injustices all around them (Marx and Engels 1975; Feuerbach 1967, 1957).

Religion is less likely to have a close relationship with power when there is official separation between church and state or when governments exercise "neutrality" by offering limited and equal support to a number of different religions (Monsma and Soper 1997). In these situations, religion is more likely to play a prophetic role. It is less likely to be co-opted by the interests of the state, serving as a check against state power. It has more flexibility to change its own internal religious structures and beliefs, and to contribute to wider social changes.

Religion and the Structure of Civil Society

The structure of civil society consists of the formal and informal relationships between state and civil society groups, and the legal or institutional measures that manage those relationships. These relationships and institutional mechanisms go some way toward determining what role religion plays in the public sphere. State/civil society relationships may be managed so that religion is—more or less—privileged, marginalized, or included on equal footing within the public sphere.

Civil society may be managed in a way that privileges religion. In these cases, religion has a closer relationship with the state than other civil society groups. This relationship might be formalized through established churches or through frequent informal contact between religious and political actors. Such relationships would be based on the assumption that religion has a unique role to play in cultivating public virtue or in building solidarity. However, religion might also fail to serve as an effective "check" on sociopolitical power. In such cases, religion would play a priestly role.

Civil society may also be managed in a way that marginalizes religion. In these cases, other civil society groups would have closer relationships with the state, and religion would be excluded from the public sphere. This situation might come about as a result of wider sociopolitical processes, such as secularization. However, it might also come about because of a policy of deliberate exclusion on the part of the state or other civil society actors. This might be the case when religion is deemed to be a private affair or to be too dangerous to be allowed in the public sphere. The exclusion or marginalization of religion might be carried out through formal mechanisms such as systematically denying religious groups funding, whilst at the same time providing funding for secular groups. Or it might be carried out through informal mechanisms, such as other civil society groups' simple reluctance to cooperate with religious actors. Or the media might be reluctant to include religious representatives in public debate. If religious actors perceive their viewpoints and rights as being systematically suppressed through formal and informal policies, religiously motivated protest or violence could result.

Finally, civil society may be managed in a way that includes religion on relatively equal footing with other civil society groups. This would include relative equality in their relationships with sociopolitical power, and in their positions in the public sphere. In such a situation, religion's contribution would not be perceived as privileged, irrelevant, or dangerous. Religion would simply be expected to compete and cooperate with other civil society groups. In such situations, religion would have the freedom and flexibility to play a prophetic role.

Religious Structures

The role of religion also depends on how different religious structures function. The following adapts Marty's (2000) analysis of religious structures (what he calls "institutions") with a view to understanding how certain structures may play priestly and/or prophetic roles. More flexible structures have greater potential to play prophetic roles. For reasons discussed below, ecumenical agencies and denominations have the least flexibility, congregations and organizations are more flexible, and networks have the greatest flexibility of all. Strong, flexible religious networks have the greatest potential to play a prophetic role.

Ecumenical Agencies

Marty (2000) says that ecumenical agencies are "one of the twentieth century's most vital and enterprising forms of advancing religious engagement with the political order" (Marty 2000:109). Indeed, it could be argued that because they bring together denominations from across the world, they are established specifically to challenge the religious status quo. The status quo had kept diverse and competing denominations from cooperating, inhibiting their ability to play a prophetic role. Ecumenical agencies have taken up a number of tasks, such as war relief, immigration activities, defending the free exercise of religion, interacting with the media, and generating debate within denominations.

However, of all the religious structures outlined here, ecumenical agencies have the most difficulty presenting a unified voice that reflects the beliefs of the people they claim to speak for. They are the preserve of high-ranking officials and theologians and are not readily accessible to the ordinary churchgoer. What ecumenical agencies say does not necessarily have a lot of support from the grass roots. This constrains their ability to play a prophetic role.

Denominations

Denominations provide congregations and the people within them with a corporate identity based on creed, confession, practical concerns, or some combination of these. Like ecumenical agencies, denominations are quite visible in the public sphere, and they explicitly address social and political issues. As such, they have the potential to articulate clear, prophetic messages. They also have the potential to serve as forums where new ideas are debated. Debates initiated by denominational elites may reach the grass roots through individual clergy, who may use denominational statements for educational purposes or to stimulate discussion. This kind of interaction and wrangling with issues can contribute to change, enhancing the denomination's potential to play a prophetic role. However, denominations represent broad constituencies and must constantly struggle to unify diversity within their ranks. This means that they may not be truly representative of their grass roots, and that internal divisions may prevent them from moving quickly to address social and political problems. This constrains their ability to play a prophetic role.

Congregations

Congregations are the places where the faithful meet on a regular basis for worship and fellowship. The interaction that takes place within congregations contributes to the wider process of socialization that occurs within civil society. It is within congregations that the faithful form their religious identities; it is also here that changes in their religious identities are most likely to be worked out. In addition, some congregations participate directly in nongovernmental politics. They may sign petitions or participate in political protests. Given their relatively small size, congregations are likely to be in touch with what is happening on the ground, allowing them to react more quickly than ecumenical agencies and denominations.

More than any other religious structure, congregations have the potential to shape the process of socialization and the formation of identity. They do so through the proclamation of religious narratives and the enactment of liturgies and ceremonies. People create bonds with each other through the intimate, face-to-face social interaction in congregations. In some cases, congregations may be relatively homogenous, and the individuals within them may have unchallenged, shared identities. However, congregations also have the potential to be places that allow for interaction amongst diverse views and identities (Marty 2000; Ammerman 2001, 1997; Ammerman et al. 1998). In such instances, congregations have the potential to drive changes in identity. Sometimes the interaction is so contentious that it produces "us-versus-them" factions, resulting in people leaving one congregation for another. At other times, interaction offers opportunities for individuals to deliberate—and perhaps to change their points of view (Marty 2000:87). Directions of change may be influenced by the leadership of clergy or by denominational statements. Local pastors provide guidance on how the members of the congregation might apply their faith to certain issues or causes. If clergy challenge long-held assumptions about the relationship between faith and politics over a sustained period of time, they may contribute to change amongst the people. Denominational statements may have a similar impact, although only a limited number of people within the congregations are likely to pay attention to denominational statements. The pastor—standing in front of them every week—is more difficult to ignore. Over time, religious identity change may lead congregations or individuals within them to play a prophetic role.

Religious Special-Interest Groups

Religious special-interest groups are established to accomplish specific goals. During this process, they also may redefine or "reframe" the way issues are discussed in the public sphere. These groups exist because the people who formed them want to instigate change. They address a seemingly limitless array of issues, including poverty relief, Third World aid, establishing prayer in schools, limiting abortion, or abolishing the death penalty. They make strident efforts to publicize their causes in the public sphere. They draw supporters from a range of denominations and claim to speak clearly on behalf of them. They are usually more responsive than denominations or ecumenical agencies, reacting quickly to changes on the ground. In particular, evangelicals are enthusiastic and efficient group-formers (Marty 2000).

Organizations are dedicated to the practice of nongovernmental politics. In a sense, religious organizations are prophetic by design. Some have narrow agendas, focusing on a few select issues. Others have broad agendas and perceive themselves as ministering to denominations or ecumenical agencies. This may include educating denominations or congregations about a variety of social or political issues, or taking the heat off denominations when it comes to advocating controversial positions (Marty 2000:142). Their influence and visibility may wax and wane, depending on the saliency of particular issues. They are dynamic and changing, constantly adjusting their message and methods to suit the situation. This makes them well-placed to play a prophetic role.

Networks

Marty does not include networks in his analysis of religious structures. This is an unfortunate oversight. A religious network is defined by the relationships that exist between various organizations, congregations, denominations, and ecumenical agencies. Religious networks are fluid and marked by a pooling of resources, personnel, and ideas. They have a greater potential than organizations to play a prophetic role because of their greater manpower, resources, and exposure in the public sphere. They may be even more effective if they are willing to cooperate with secular allies (Appleby 2000; Johnston 2003). In some contexts, religious networks may contribute to social movements that bring about widespread change.

More than any other religious structure, networks have the potential to shape the process of socialization (including identity change) while at the same time participating effectively in nongovernmental politics (including reframing sociopolitical projects). Networks are places where interpersonal relationships are formed as people exchange resources and ideas. This can lead to a questioning of accepted identities and political frameworks, and to the gradual formulation of alternative identities and frameworks. In networks, this occurs through the quiet collaboration between organizations, congregations, denominations, and ecumenical agencies on teaching or research initiatives, or in joint worship services or fellowship groups.

Networks, like organizations, are dedicated to the practice of nongovernmental politics. However, their greater humanpower and resources distinguish them from organizations. They harness the energy, brainpower, and resources of the enthusiastic activists from a variety of religious structures. This gives them visibility in the public sphere and greater opportunity to argue for change. Although a network will not last indefinitely, it can contribute by way of concrete reforms, reframing debates, garnering recognition for new identities, and mobilizing people to participate in the democratic process. However, since networks may include people with broad and differing points of view, they may have trouble maintaining unity. Religion's prophetic potential is maximized when there are unified, well-developed religious networks.

Religious Belief

When people of faith participate in the public sphere, their actions are enabled or constrained by the structures outlined above. However, people are not structural puppets. It matters *what* people believe. Religious ideas inform their perceptions of what the sociopolitical order *should* be like. Religious ideas shape people's perceptions of the "proper" relationship between religion and the state, and of the way that religion should influence society and culture. These ideas provide people with the motivation to act to achieve their visions.

It becomes more complicated for people to achieve their religious, political, or social visions in divided societies. In divided societies, the religious beliefs that matter most are those about the relationship between church and state, about religious or cultural pluralism, and about violence and peace. The way particular theological traditions handle these issues affects the ability of religion to contribute to conflict or peace. For example, religious traditions that advocate a close relationship between church and state, such as some forms of covenantal Calvinism, may be less open to tolerating religious and cultural pluralism within the wider society. Traditions that advocate a strict separation between church and state, such as Anabaptism, may be more open to tolerating religious and cultural pluralism. The task then becomes to analyze how those traditions fit within plural, divided societies and to strengthen the aspects of those traditions that advocate peaceful, pluralist coexistence.

Religious attitudes toward violence and peace are also crucial. In a comparative study of religion and conflict, Appleby (2000) argues that the content of religious beliefs is decisive when it comes down to whether religion will contribute to conflict or peacemaking. In contexts where the population is religiously "illiterate," people may know their holy book or engage in religious practices, but they have a low level of knowledge about or do not reflect on the peacemaking tenets of their religion (Appleby 2000:69). They are vulnerable to demagogues who exploit religious symbolism for sinister purposes. The result may be fundamentalism, religious nationalism, and violence. On the other hand, when the population is religiously "literate," people have a deep knowledge of doctrinal and moral teachings, especially the beliefs and practices of the religion that "reinforce and contextualize the priority given to peace and reconciliation" (Appleby 2000:119). The result is the emergence of nonviolent "militants for peace." The task then becomes identifying and strengthening those aspects of religious traditions that focus on peace and reconciliation.

Civil Society in Northern Ireland

In Northern Ireland, there has been some use of the conceptual tools discussed above. These have been employed both by peace and conflict resolution organizations (PCROs) and by the British government. The following evaluates how some conflict management and transformation strategies have been employed. It is concerned with equality, inclusion, and exclusion within civil society—especially the extent that evangelicals (and other religious actors) are included. Previous research on the contribution of religious actors in Northern Ireland fails to give a coherent or systematic account of the activities of evangelicals. These considerations lay the groundwork for the analysis of evangelicalism presented in the following chapters.

The Role of PCROs

The focus of much of the Northern Irish research on civil society has been on PCROs.[6] This research has praised PCROs for their contributions and credited them with pushing elites toward accommodation.[7] Wilson and Tyrrell (1995) claim "the role of mediation projects and reconciliation agencies is to build a culture of dialogue in a divided society where the pull is to split into exclusive 'tribes' at times of crisis" (Wilson and Tyrrell 1995:246). They argue that the role of PCROs has been reflected in changes in government policy, including the introduction of Education for Mutual Understanding (EMU), the creation of the Community Relations Council, the introduction of Community Relations Officers in District Council areas, the promotion of cross-community work in schools, and the formulation of antidiscrimination policies of the trade union movement Counteract. They also argue that the impact of PCROs could be felt in areas such as education, youth work, and community development. Brewer (2003a) says that PCROs have aided the peace process by filling in the "democratic deficit" created during Direct Rule, when the accountability of the state was diminished and local politicians failed to deal with "bread and butter" concerns. Some groups have empowered women (see also Brewer, Bishop, and Higgins 2001), begun to create a shared, cross-community identity, and developed "a form of communitarianism" and a "local civic culture in face of the divided society surrounding them" (Brewer, Bishop, and Higgins 2001:132). Similarly, Cochrane and Dunn credit PCROs with creating an "NGO culture," which includes

> a belief in community rather than individualism; a concern to work for the betterment of that perceived community . . . a sense of moral imperative to make a positive contribution to that defined community; an optimism that such activity will 'make a difference'; the strength of personality to run an organization on very limited resources and often against opposition from one (or both) of the main communities in Northern Ireland; a set of moral values taken from the NGO sector which provide many PCROs with an organizational ethos based on concepts such as non-violence, citizenship, dialogic democracy, pluralism, multiculturalism and partnership. (2002:97–98)

In their attempt to gauge the influence of PCROs on the peace process, Cochrane and Dunn (2002) admit that measuring the effects of PCROs quantitatively is "impossible." They draw some qualitative conclusions based on a wide range of interviews with critics and advocates of the PCRO sector. Among the identified "successes" of the PCRO sector are the growth of integrated education, the development of a culture of dialogue, the Opsahl Commission,[8] the empowerment of community activists who would later play a key role in political negotiations,[9] and the promotion of the Belfast Agreement during the referendum campaign. They argue that grand, highly publicized gestures such as peace marches have had little lasting effect, whereas grassroots work carried out by "unsung heroes" has made a more lasting contribution.

However, critics argue that this PCRO picture of the world is wildly unrepresentative of what civil society is really like. This assessment contributes to their conclusion that civil society is either a marginal player or not important at all in conflict management (McGarry and O'Leary 2004). Focusing exclusively on the contributions of PCROs makes it easy to forget that the majority of civil society may not share a cross-community, dialogic ethos at all. This falls into the trap of defining civil society in terms of "good" PCROs. The tendency to present PCROs as "heroes" cajoling unwilling political elites to reach a settlement also obscures the extent to which government can play a role in restructuring civil society so that it is conducive to transformation rather than to conflict.

The British Government's "Civil Society Approach"

Some PCROs have addressed the socialization of people into opposing identities, attempting to construct shared identities and a dialogic ethos. However, the extent that PCROs have acted independently has been exaggerated. It is often not recognized that the British government's attempts at conflict resolution have included a "civil society approach" (Dixon 1997). This approach, though often unsystematic and ad hoc, reflects the assumption that the conflict is, at least in part, one between oppositional identities. It has included many of the reforms advocated by multicultural theorists. This approach has been implemented concurrently with political negotiations and changes in governmental institutions and is complementary to that process (Dixon 2005; Farrington 2006, 2004b; Bloomfield 1996).

The civil society approach has been part of British government policy in Northern Ireland since the early 1970s. It should be understood in two contexts: as part of an overall strategy to equalize social, cultural, and political conditions between Catholics and Protestants; and as a reflection of changes and emphases in the wider, UK approach to civil society. For example, throughout the Troubles, the British government introduced a series of fair employment legislation that has gone some way toward redressing economic discrimination.[10] It also introduced a number of reforms designed to manage civil society by encouraging good relations through cross-community contact (Dixon 1997; Fitzduff 2002). Although these broad reforms have been important in preparing the ground for political initiatives and in redefining interests, alliances, and identities, this research will not discuss them in-depth.[11] Rather, the "narrower" aspects of the government's approach are most relevant to this study, as they have more immediate impacts on the restructuring of civil society. For example, they provide civil society actors with incentives to change their identities and/or their sociopolitical projects to fit within new, emerging civil society structures.

Some of these more narrow reforms have been influenced by wider UK and European trends toward partnership, governance, and the rhetoric of dialogic democracy (Kearney and Williamson 2001; Acheson et al. 2004). These trends intensified after the election of a Labour government in 1997. In particular, Prime Minister Tony Blair was an enthusiastic proponent of the "Third Way" ideas promoted by sociologist Anthony Giddens (Farrington 2004b; Bacon 2003; see Giddens 1998). Giddens argued that state and civil society should work closely together, with the state supporting, funding, and consulting civil society. The EU often made its funding contingent on civil society groups adopting European norms of social partnership and dialogue (Brewer 2003a; Taylor 2001).[12]

The Community Relations Council

The CRC (established in 1990) was created specifically to address the limitations of civil society in a divided society. It draws on themes explored by multiculturalists such as Kymlicka and Parekh, including promotion of primary and shared identities. It developed out of the Central Community Relations Unit (CCRU) within the Northern Ireland Office, which was set up in 1988 to promote conflict resolution work (Fitzduff 2002). The CRC claims to be an independent body, but it works closely with the government and receives funding from the government and the EU. The training, funding, and oversight roles of the CRC have been expanded following consultation on the government's "Shared Future" document (2005).

The CRC is the main distributor of funds for civil society groups. Its funding and training programs emphasize cross-community work. This is based on the premise that cross-community contact encourages conflict resolution (Varshney 2003, 2001; Bloomfield 1996). As Bloomfield concludes, it is likely that groups that participate in training schemes are better equipped to apply for and receive CRC funding: "Whether wholly intentional or not, the CRC's control of funding gives it the muscle to encourage its client groups to follow the direction and policy it wants them to" (1996:154).

The CRC's attempts to introduce community relations to "other social groupings" have been manifested in its support for "single-identity" work. The logic behind single-identity work is that it should eventually lead to cross-community work. This reflects the arguments of analysts such as Kymlicka, who says that primary identities must be strengthened before people can come to hold a shared identity or to value "deep diversity." This is clear from the CRC's definition of single identity:

> Single identity work is a term sometimes used to describe activities which do not cross traditional community boundaries. It aims to increase confidence within a community so that people are better able to define their identity and needs in relation to others. Community

relations work of this type should challenge long-held, unquestioned stereotypes which may no longer fit within that community and should open up channels of communication within communities and between communities. ("Community Relations: A Brief Guide," n.d.)

The main criticism of single-identity work has been that it "essential-izes" differences, entrenching opposing identities and making it more difficult for radical change to take place. For instance, Cochrane and Dunn reported that single-identity work "tends to reinforce ethnic identity, fuel a victim-culture, and corral communities behind their tribal fences, rather than diminish stereotypes or increase intercom-munal understanding and toleration" (2002:169). On the other hand, it has been argued that "in some circumstances, and within certain contexts, single-identity work was the only realistic peace/conflict resolution strategy available" (Cochrane and Dunn 2002:169–170). The ability of single-identity work to promote conflict resolution remains unproven.

As is the case with PCROs, it is difficult to evaluate the effective-ness of the CRC in achieving its goal of promoting good community relations. Bloomfield concluded that the "single most important effect" of the CRC's work has been "widening the arena of commu-nity relations" so that its assumptions and discourses have begun to permeate the wider society (1996:167). At the least, the British gov-ernment has expressed the desire to continue funding and developing civil society groups that promote a cross-community, dialogic ethos. As the "Shared Future" document states, this includes plans to make the CRC "more broadly representative of civic society to enhance its funding powers; and to give it the task of evaluating the community relations programs of civil society groups and District Councils" ("A Shared Future" 2005).[13]

Mainstreaming Third Way Governance

The British government increasingly has moved to define government–civil society relationships. This reflects an attempt to mainstream Third Way-style governance in Northern Ireland. This has created an officially sanctioned public sphere in which govern-ment has increasingly attempted to lay down rules and regulations for the participation of civil society groups. As in the wider UK, Third Way governance is concerned with involving civil society in perform-ing various social or political functions previously performed by the state. In Northern Ireland, however, governance includes an emphasis on cross-community interaction and parity of esteem for competing identities. To some extent, funding and favor for civil society groups are dependent upon their acceptance of this approach.

This process can be analyzed in a series of documents, discussion, and consultation papers published since the early 1990s. The 1993 document "Strategy for Support of the Voluntary Sector and for Community Development in Northern Ireland" formalized relationships between the state and civil society. It was followed by the 1998 document "Compact: Between Government and the Voluntary and Community Sector in Northern Ireland—Building Real Partnerships." In 1998 government established the Joint Government Voluntary and Community Sector Forum for Northern Ireland in an effort to create "a formal mechanism for promoting dialogue between Government and the sector" (Kearney and Williamson 2001:60). After the Belfast Agreement, ministers in the Northern Ireland Executive and MLAs continued to develop a governance approach. The Northern Ireland Executive's first programme for government, "Making a Difference 2002–2005" (2001), explicitly recognized the value of the voluntary and community sector. The document "Partners for Change: Government's Strategy for Support of the Voluntary and Community Sector 2001–2004," produced by government and community representatives from the Joint Government Voluntary and Community Sector Forum, committed the government to three-year plans of practical action to involve civil society in policy development, established mechanisms to monitor the input of civil society, and provided civil society with "comprehensive information on Government support and plans for the sector" (Kearney and Williamson 2001:61). Kearney and Williamson described this as a "pioneering" document "without precedent in the UK" (2001:68). Another document, "Building on Progress: Priorities and Plans for 2003–2006," identified specific government commitments to the sector.

The 2003–2005 "Shared Future" consultation process has attempted to formalize relationships between government and civil society and establish clear guidelines about funding, expectations, and how to evaluate the effectiveness of civil society groups. The original consultation paper was released in 2003 and invited organizations and individuals to comment on the policy outlined in it. It was sent to "MPs, MLAs, key statutory agencies and a wide range of voluntary and community organizations" and was also posted online ("A Shared Future," 2003). A series of discussion forums, many of which were facilitated by the CRC, were held. The consultation process lasted until September 30, 2003. The process "engaged more than 10,000 people and generated over 500 written responses from across society" ("A Shared Future" 2005).[14] At the conclusion of the process, government announced that it would work with civil society representatives to develop triennial action plans, that the roles and functions of the CRC would be enhanced, and that from 2007 a new District Council Good Relations Challenge Programme would be established ("A Shared Future" 2005).

At the same time, the Department of Social Development (DSD) established a taskforce of five statutory representatives and six independent experts to research best practices for funding the sector, particularly in light of decreased funding from outside sources in the new "postconflict" phase.[15] The taskforce report recommended that government provide more stable, efficient, and long-term funding sources for the voluntary sector, the establishment of a Community Investment Fund for the support of local community development, and the establishment of a Modernisation Fund to enhance the ability of community organizations to provide public services ("Investing Together" 2004).

Increasingly formalized government support for the sector can be viewed as a welcome safeguard in that funding, in particular, may help some groups. On the other hand, the ongoing development of government–civil society relationships defines who is "in" and who is "out" in terms of access to government and funding. For instance, the responses to the Shared Future document were broadly supportive of the civil society approach and of efforts to build a nonsectarian society based on strong cross-community relationships. But most of these responses came from groups already committed to a cross-community approach.[16] This indicates that the consultation process did not stimulate a great deal of debate with those groups that do not already enjoy mutually supportive relationships with government. Groups that toe the government line have greater access to and interaction with government. Not only does this compromise the ability of some groups to maintain their independence, it systematically excludes and alienates other groups. This may particularly be so in the case of churches or faith-based organizations, which may have more difficulty than other groups in meeting cross-community or equality requirements.

The Civic Forum

The Belfast Agreement created new political institutions that changed the dynamics of the relationships between government and civil society. The Civic Forum was the most obvious and visible attempt to mainstream governance and to continue with a cross-community, civil society–based approach to conflict management. Its role was outlined in section 34 of strand one of the agreement:

> A consultative Civic Forum will be established. It will comprise representatives of the business, trade union and voluntary sectors, and such other sectors as agreed by the First Minister and Deputy First Minister. It will act as a consultative mechanism on social, economic and cultural issues. The First Minister and Deputy First Minister will by agreement provide administrative support for the Civic Forum and establish guidelines for the selection of representatives to the Civic Forum.[17]

The Civic Forum struggled to work out its role within Northern Ireland. Neither the Belfast Agreement nor the Office of the First Minister and Deputy First Minister (OFMDFM) provided much direction in this regard. Rather, members of the forum were left to their own devices in defining and carrying out its tasks (Palshaugen 2004, 2002). They decided that the key priorities were providing advice to government, responding to consultation papers, and consulting widely with civil society and key sector groups. Before it was suspended along with the Assembly in 2002, the Civic Forum was involved in the development of the Programme for Government and produced three reports: "Can Do Better: Educational Disadvantage in the Context of Lifelong Learning" (September 2002), "Education Disadvantage: A Civic Discussion" (April 2002), and "A Regional Strategy for Social Inclusion" (September 2002). The Civic Forum did not receive maintenance funding when it was suspended, unlike the Assembly and the North-South institutions.

Palshaugen's study of the Civic Forum conveys the sense in which members, while believing in the potential of the forum, had concerns about what the forum was actually doing (or not doing). Members felt that politicians in the Assembly did not take their work seriously. No formal or consistent working relationship was established with the Assembly or the OFMDFM. On the other hand, some felt that the forum was too compromised in its relationship with the OFMDFM and could not be truly independent. The forum also had little communication with the grass roots and had no strategy of communicating its work and its recommendations to the wider public through the media or by other means. Farrington (2004b) notes that almost no one agreed on what the Civic Forum should be doing. Ultimately it did nothing as civil society activists pursued other avenues of access and influence (Farrington 2004b:7–8). The work that it might have been doing has been taken up by two alternative forums—an economic development forum and the Joint Government Voluntary and Community Sector Forum for Northern Ireland. These forums and informal contacts between government and civil society actors are the avenues through which government–civil society interaction have been taking place (Farrington 2004b). Although the Civic Forum has not lived up to its potential and seems unlikely to be revived, other avenues for government–civil society interaction have emerged.

Limits of the Civil Society Approach

The British government's civil society approach has reflected the assumptions of multicultural theorists and has adopted some of their strategies for managing conflict. Specific strategies such as the CRC, mainstreaming Third Way governance, and the Civic Forum have met with mixed results. Critics claim that the grass roots have been encouraged to develop divisive "identity politics" that essentialize differences, rather than developing a shared identity. They cite evidence that attitudes on core issues such as identity, the British and Irish states, and support for the Belfast Agreement remain polarized and that housing is becoming more segregated (Hayes and McAllister 1999). There has been no significant move toward integrated education.

But if these strategies have not fulfilled their aims, some potentially useful strategies have not been considered sufficiently or at all. For instance, the British government has not been systematic in its attempts to include victims in the public sphere or to construct anything resembling Keane's (1998) "public spheres of controversy" for monitoring violence. It appointed a Victims Commissioner in 2005, but it is not yet clear how effective the commissioner's work has been. Government-initiated discussions about establishing a truth commission similar to the Truth and Reconciliation Commission in South Africa were deemed impractical and were unpopular, especially with unionists (Guelke 2004). The British government's Bloody Sunday inquiry and proposed inquiries into the murders of Rosemary Nelson, Billy Wright, Robert Hamill, and Pat Finucane generated bitter debate. Unionists felt that these inquiries focused disproportionately on violence carried out by representatives of the British state or loyalists, whilst downplaying the violent acts of the Irish Republican Army. Nationalists felt that the stipulations of the inquiries were not transparent enough and would allow the British state to prevent the "truth" from emerging in the public sphere. In 2007, Northern Ireland Secretary Peter Hain appointed a consultative group to determine how to deal with the legacy of the Troubles.[18] A critique of the British government's ad hoc and failed attempts to include victims in the public sphere is beyond the scope of this research.[19]

Of immediate concern to this research is the failure of the British government's civil society approach to include religious groups in general—and evangelical groups in particular—in the public sphere. This is striking, especially since the Labour government has been keen to include faith-based organizations in its plans for Third Way governance in other parts of the UK (Bacon 2003). The reluctance to include religion may reflect sentiments such as those of Little (2004), who argues that religion in Northern Ireland is so divisive that it *should* be excluded from the public sphere. This is reflected in the "definition of the voluntary sector followed by the NI [Northern Ireland] Community and Voluntary Sector Almanac," which "virtually excludes churches, asserting that they exist 'solely for the benefit of their members' (NICVA 2002, p. 13)" (Bacon 2003:27).

That said, there are signs that there have been efforts to include religious perspectives. Representatives from the churches were included in the Civic Forum. Power (2005) has documented how the CRC has supported the establishment of "Church Fora," which are cross-community church groups that address "social and community needs, reconciliation issues and community life" (CRC definition of Church Fora, quoted in Power 2005:59). Bacon's research (2003, 1998) attempts to gauge the inclusion of religious actors in the public sphere. Although Bacon observes that the Northern Ireland Executive's 2001 Programme for Government "did not go out of its way to embrace faith communities" (2003:26), faith communities were singled out for emphasis during the "Shared Future" consultation. However, it is not clear what *kind* of religion government and other civil society actors are willing to include within the public sphere. The most that can probably be concluded is that in Northern Ireland there is openness to including "nice" religious activists—especially those that agree with the government's civil society approach. This is the case with a group such as Evangelical Contribution on Northern Ireland (ECONI), which has received both core funding and grants from the CRC.[20]

Conclusions

This chapter has provided conceptual tools for thinking about civil society, religion, and conflict in divided societies. These tools can help to understand how civil society functions, religion's role within it, and how it can be an arena both for conflict and transformation. In conflicts with religious dimensions, the role of religion must be understood and should be incorporated into conflict management and transformation strategies. In Northern Ireland, the conflict management strategies employed have met with, at best, mixed results. It is easy to fall into the analytical trap of cataloguing the positive and negative effects of particular strategies, such as the CRC or Civic Forum. While it is important to analyze concrete examples of "success" or "failure" brought about through these strategies, what is more significant is the way in which they have changed the *structure* of Northern Irish civil society. Now, the structure is such that civil society groups that have the "right" kind of single identity or that participate in cross-community dialogue receive government funding and favor. Groups that do not approve of the government's civil society approach face a comprehensive network of institutions that are meant to enforce these norms, such as the CRC. Groups that do not approve are excluded if they do not sign up to these norms. This creates a powerful structural incentive for these groups to protest (violently or nonviolently) against these norms or to change to fit in with them. Religious actors must deal with these structures as they attempt to work out their place in the new civil society that is emerging.

Chapter 3

Religion in Transition—Comparative Perspectives

This chapter explores the role of evangelicalism in Northern Ireland over time and from comparative perspectives. In the first part of the chapter, concepts about the role of religion outlined in chapter 2 provide a framework for understanding developments and change, including the relationship between evangelicalism and power, evangelical responses to changing social and political structures, and the content of evangelical beliefs. These concepts are used to explain the historical role of evangelicalism, as well as evangelicals' responses to changes since the imposition of Direct Rule in 1972. A key argument is that changes since 1972 have seen evangelicalism lose a privileged relationship with social and political power, and traditional and mediating evangelicals have adapted in different ways. Their ability to adapt is constrained by the changed structures of civil society (chapter 2) and by religious structures in Northern Ireland.

In the latter part of the chapter, comparative perspectives are employed to help understand how Northern Irish evangelicals are negotiating the breakdown of their relationship with power. Analyzing two contexts in which evangelicalism once held a privileged position—the USA and Canada—allows for examination of how new forms of evangelical activism emerged there. These forms were shaped by different theological emphases and sociostructural conditions. For example, in the USA, evangelical activism is dominated by a vocal "religious right" that focuses on moral issues and is closely aligned with the Republican Party. Evangelicals' strategy of engaging in interest-group politics gives them scope for playing a prophetic role and contributing to social and political change. This style is well-suited to American political structures, which encourage interest-group activity. However, much evangelical argument continues to be laced with "Christian America" themes that reflect older Calvinist conceptions of the relationship between church and state, and marks Americans out as a chosen people. This line of argument contributes to a "culture war" atmosphere that may adversely impact evangelicals' ability to interact with other civil society groups (Hunter 1991).

In Canada, on the other hand, evangelical activism is less aligned with either the right or the left or a particular political party. Canadian evangelicals have developed what Noll calls a "mediating" form of activism, in which they cooperate willingly with other groups in civil society. This strategy gives them scope for playing a prophetic role and for working with other groups for social and political change. Canadian evangelicals do not see the development of a pluralist society as unfortunate nor do they draw on Calvinism to argue for a Christian Canada. This style is well-suited to Canadian political structures, which do not encourage interest-group activity but rather encourage consensus building amongst civil society groups. This means that although Canadian evangelicals are usually less noticeable in the public sphere than their American counterparts, they may still play a prophetic role.

In Northern Ireland, some similarities in history and contemporary trends indicate that evangelicalism may move in an American direction; in other ways Northern Irish evangelicalism seems more like its Canadian counterpart. The benefits for conflict transformation of the Canadian model of mediation and consensus building are obvious. But even movement in the American direction, in which evangelicals adapt critically and selectively to the new sociopolitical institutions, is more promising for conflict transformation than if evangelicals outright rejected the principles behind the Belfast Agreement, such as power sharing.

Evangelicalism in Northern Ireland

Evangelicalism and Power in Historical Perspective

Evangelicalism in contemporary Northern Ireland cannot be understood without understanding the history of evangelicalism in Ireland, especially its relationship with social and political power. In the mid-eighteenth century, evangelicalism found fertile ground for expansion amongst Protestants in Ireland's northernmost province of Ulster. The bearers of the evangelical message encountered a Protestantism that was divided amongst adherents to the established (Anglican) Church of Ireland, Presbyterians, Covenanters, and other small groups. Those of Reformed, Calvinist traditions were a clear numerical majority. There were significant tensions among these different Protestant groups, and for a time, Presbyterians and other "dissenters" faced official discrimination under the penal laws. However, most Protestants and dissenters occupied higher socioeconomic positions than the native Irish Catholics and they were united in opposition to the Catholic religion. This opposition was both theological and political: it was believed that the doctrines of the Catholic Church were simply wrong and that the authoritarian structure of the church prevented its members from becoming citizens committed to liberty and freedom. Protestants were also wary of the overwhelming Catholic majority on the island (Hempton and Hill 1992; Thomson 2002; McBride 1998; Brewer and Higgins 1998; Miller 1978).

Adherents to the varieties of Protestantism tended to put aside their differences during periods of social and political stress (Ruane and Todd 1996; Bruce 1986). Evangelicalism, as a transdenominational movement, increasingly provided Protestants with a unifying source of identity at these times. In the 1830s, for instance, Daniel O'Connell's Catholic Association issued a direct challenge to Protestant socioeconomic privilege. The Catholic Association created alliances between clergy and Catholic laypeople for the goals of emancipating and removing/reducing the taxes that were paid to the established Church of Ireland. O'Connell and his allies were eventually successful in pressing the Westminster government for reforms. In response, Rev. Henry Cooke, a Presbyterian, argued "that state support for a Protestant church, even if it was the wrong Protestant church, was better than a free market system in which Roman Catholicism would come to be dominant by virtue of numbers" (Bruce 1986:5; Thomson 2002). The Orange Order emerged and consolidated its influence during this time.[1] Though originally linked with the Church of Ireland, the Orange Order unified diverse expressions of Protestantism "into a loose coalition with the religiopolitical aim of protecting Protestant faith and power" (Mitchel 2003:135). Its religious ethos gradually became more evangelical. The boundaries had begun to harden between Catholicism and a Protestantism that was becoming ever more united behind Calvinist-inspired evangelicalism.

The Great Irish Famine, which disproportionately affected the largely Catholic west of Ireland, further contributed to the hardening of boundaries. It was easy for some Protestants to interpret the famine as God's judgment on "wrong religion"—Catholicism. This interpretation of events could be supported by the observation that Ulster, where most of the Protestants lived, was relatively untouched by the famine.[2] In addition, there was a massive evangelical revival in 1859 (Thomson 2002; Wells and Livingstone 1999). The revival was a local manifestation of the Great Awakenings in the USA and England. In Ulster, it had social effects such as increased Sabbath observance, the closing of pubs, and a decrease in gambling. Few Catholics were affected by the revival, which tended to confirm Protestants' percep- tions that their religion was just wrong. This, coupled with the Catholic revival under Cardinal Paul Cullen in the 1860s, contributed even further to the building of barriers between Protestants and Catholics (Liechty and Clegg 2001; Hempton and Hill 1992).

By the late nineteenth century, then, evangelicalism informed Protestants' social mores and political discourses, and evangelical assumptions were articulated at both elite and populist levels. This was reflected in the Ulster Solemn League and Covenant of 1912, the purpose of which was to proclaim Protestant determination to resist "home rule" for an Irish parliament. It drew Scottish Calvinist and Covenanting traditions under the "sacred canopy" of evangelicalism (Brewer and Higgins 1998). The Solemn League and Covenant iden- tified evangelicalism with the sociopolitical power of the Protestants of Ulster.

The home rule issue was set aside during World War I. But after the Irish War for Independence (1919–1921), Ireland was partitioned: the 26 counties of the South became the Irish Free State, while six of the northern counties remained within the United Kingdom. With Protestants in a numerical majority in Northern Ireland, unionists set about building a Protestant parliament for a Protestant people.[3] Bruce (1986) calls this period "a time of considerable Protestant unity" and argues that the importance of Presbyterianism increased at this time. Evangelist A.P. Nicholson sparked a series of revivals in the 1920s, contributing to the process whereby Protestants "[rebuilt] their sense of community and reaffirm[ed] those values which they hold central to their identity" (Bruce 1986:15).

Mitchel calls this period, which stretched from 1921 until the imposition of Direct Rule in 1972, the "golden era" of Ulster union- ism (2003:86). The Orange Order understood itself as a bulwark against "absorption within an all-Ireland Catholic state, a fate con- trary to God's will" (Mitchel 2003:159–160) and dismissed Catholic grievances, arguing that good citizenship meant supporting the union and Protestantism. The largely evangelical (but divided) Presbyterian Church also supported unionism uncritically and participated in the hardening of boundaries between Catholic and Protestant (Mitchel 2003:237). Evangelicalism played a "priestly" role for Ulster's Protestants, underwriting the assumptions of unionism and maintaining its privileged position. It supported the state and its social and political structures.

The Content of Ulster Evangelical Beliefs

Evangelicalism does not have a unified theology or an agreed-upon position on the beliefs that matter most in divided societies: the relationship between church and state, religious and cultural pluralism, and violence and peace. Rather, evangelicals tend to "borrow" the beliefs that are dominant in the Christian denominations in their host culture. In Ulster, evangelicals borrowed from the covenantal Calvinism that was dominant in the eighteenth century. The concepts and symbols of Calvinism infused the Ulster evangelical ethos and provided the resources for a "Protestant ideology" (Wright 1973).

Calvinism's conception of a covenantal relationship between church and state shaped evangelical attitudes about the Catholic religion and the ability of Catholics to be "good" citizens. Covenantal Calvinism strives to order church-state covenants to reflect God's laws. It is believed that if the state follows God's laws, God will bless it. If it disobeys God's laws, God will curse it. For Calvin himself, that meant establishing the "Christian Commonwealth" of Geneva, where all citizens covenanted to uphold the Ten Commandments. For Cromwell's English Puritans or the Puritans of New England, that meant a coercive Christian commonwealth. For the Scots who established the Solemn League and Covenant of 1643 with the English Parliament, that meant eliminating Catholicism in Scotland and extending Presbyterianism to England and Ireland. For covenantal Calvinists, it follows that some religions are simply wrong and that allowing them free reign can wreak havoc in the body politic. In evangelical Ulster, this provided a powerful theological justification for sociopolitical power. Protestants—who had the right religion— must maintain their sociopolitical power in order to ensure God's blessing.

The covenant also shaped evangelical attitudes about religious and cultural pluralism. Catholics, because their religion was "wrong," simply could not uphold the covenant. This created a sharp contrast between Protestants and Catholics. Protestants were a "chosen people," dutifully upholding God's laws and bringing order and civilization to the land. Catholics were outside the pale, "barbaric and uncivilized."

In evangelical Ulster, this provided a powerful theological justification for resisting integration with the Catholic Irish.

Finally, the covenant provided justification for violence. The covenant required faithful citizens to monitor the state. It followed from the covenant that if the state was not fulfilling its part of the covenant, then the Christian citizens living within it could resort to legal (or constitutional) agitation. If that failed, violent revolution could be justified. In evangelical Ulster, this has provided a powerful theological justification for resistance to the policies of the British crown or the British state. When Protestants threatened to resort to armed insurrection if their opposition to home rule was not heeded, they followed the logic of covenantal Calvinist thinking.

These Calvinist ideas shaped evangelicals' perceptions of what the sociopolitical order should be like. They justified Protestants' privileged relationship with sociopolitical power, and they legitimated their ("divine") right to rule. They underpinned Protestants' resistance to integration with Catholics, contributing to an uneasy religious and cultural pluralism on the island. And they provided Protestants with a justification to be violent or to threaten violence, even if it was as a last resort.[4]

Evangelicalism and Change in Northern Ireland

Northern Ireland during the Stormont era (1921–1972) was "peaceful" in that there was little sectarian violence.[5] The civil rights movement (1964–1969) and the start of the Troubles changed this state of affairs. The subsequent breakdown of Protestant power created a space in which evangelicalism could play a prophetic rather than a priestly role.

Changes from 1972

It is difficult to overestimate the extent of the changes in power relations, and in the structure of the Northern Ireland conflict, since 1972. From 1921 to 1972, the British government was reluctant to intervene in Northern Ireland's internal affairs, which allowed unionists to govern with a relatively free hand. At this stage, the structure of power relations in Northern Ireland was "triangular," with the British state and the two communities comprising the three main actors. This structure was marked by unequal power relationships between the communities and in the communities' relationships to the British state (Ruane and Todd 1996). However, in Northern Ireland, the unrest caused by the civil rights movement and the consequent outbreak of violence, followed by the deployment of British troops on the streets, and the institution of Direct Rule in 1972 unsettled the balance of power. This resulted in a push for the changes in power relationships to be reflected in political institutions. Accordingly, the British government endorsed the idea of a power-sharing Assembly and a north-south Council of Ireland that would recognize a role for the Irish State in Northern Ireland. These reforms were reflected in the Sunningdale Agreement of 1973, which was negotiated by the British and Irish governments and the three pro-power-sharing parties: the nationalist Social and Democratic Labour Party (SDLP), the Ulster Unionist Party (UUP), and the Alliance Party (Dixon 2001b: 129–157). These negotiations excluded the representatives of a substantial portion of the unionist electorate, including Ian Paisley's Democratic Unionist Party (DUP) and the Vanguard Unionist Progressive Party (VUPP) both of which were bitterly opposed to the reforms, especially the Council of Ireland.[6] In the midst of growing unionist discontent, the loyalist Ulster Workers' Council (UWC) strike of May 1974 brought down the power-sharing executive.

However, the UWC strike was not a clear-cut victory for unionists and there would be no return to a pre-1972 institutional arrangement. From Sunningdale to the Anglo-Irish Agreement in 1985, attempts at reform included power-sharing and binational (British and Irish state) dimensions—even under the watch of Margaret Thatcher, who was regarded as sympathetic to unionists (Dixon 2001a; Cunningham 1991, Moloney 2002). The Anglo-Irish Agreement, which was negotiated by the British and Irish governments, included a British-Irish intergovernmental conference dealing with security, legal/justice issues, cross-border cooperation, identities, cultural traditions, and discrimination; a permanent secretariat of British and Irish civil servants; and provisions for a devolved, power-sharing Assembly (Dixon 2001b: 190–214). Unionists perceived it as a step closer to a united Ireland and believed that the British government had negotiated it over their heads, whilst the Irish government had represented nationalists' interests. This prompted widespread unionist disaffection, including a sustained campaign of civil disobedience, protests of up to 200,000 demonstrators in 1985 and 1986, and a loyalist strike in 1986 in which 47 Royal Ulster Constabulary (RUC) officers were injured (Dixon 2001b:205–210; Farrington 2006). These protests demonstrated the extent of unionist discontent with the agreement, but unlike Sunningdale, they did not bring it down. The Anglo-Irish Agreement provided a more durable institutional structure for reflecting the changing balance of power in the conflict. Rather than a triangular form, the conflict had moved to a "symmetrical" form in which the major players were the two communities and the British and Irish states (Ruane and Todd 1996).

By this stage, power-sharing and binational dimensions had achieved a "nonnegotiable" status—it was generally accepted that they must be included in any future settlement. Moreover, British government reforms intended to reduce social, economic, and cultural inequality between Protestants and Catholics had begun to take effect, further reducing the power gap between the two communities. The provisions of the Belfast Agreement reflected the change from a "triangular" to a "symmetrical" structure of conflict and included a power-sharing Assembly and a north-south body—the North-South Ministerial Council (Ruane and Todd 1996).[7] It also included east-west bodies and provisions for human rights, equality, and the recognition of identities, making it Sunningdale-plus rather than "Sunningdale for slow learners" (Dixon 2001a).[8]

Clearly, there would be no going back to a Protestant parliament for a Protestant people. The social, economic, cultural, and political gaps between Protestants and Catholics had narrowed, and Protestants no longer retained a privileged position with the state. Indeed, Protestants usually felt abandoned by the British state and viewed it as a dubious ally. This was a context in which evangelicalism could no longer play a priestly role. Evangelicalism could not legitimate and underwrite a form of Protestant power that no longer existed.

Evangelical Responses to Changes

One evangelical response to the changes was the "revival" associated with Rev. Ian Paisley's street politics and the rapid growth of his Free Presbyterian Church.[9] Paisley also founded the DUP, originally called the Protestant Unionist Party, during this time (Bruce 1986; Smyth 1987; Moloney and Pollak 1986; Cooke 1996; O'Callaghan and O'Donnell 2006). Paisley is a controversial and charismatic figure who continues to lead both his church and his party, claiming that he holds fast to traditional evangelical positions and that he fights "for God and Ulster" (Southern 2005).

Another response was the formation of the political action group Evangelical Contribution on Northern Ireland (ECONI) in 1985. The founders of ECONI were reacting in part out of distaste at what they perceived as Paisley's inappropriate mixing of religion and politics. They repudiated covenantal Calvinism and attempted to construct an evangelical theology that drew largely on concepts from Anabaptism. These two broad responses were very different in style and content. But they were both, in their own way, prophetic. They challenged the way things were and the way things were going.

These evangelical voices are difficult to ignore, even if religion's critics (especially if they associate evangelicalism with Paisley) have been keen to keep them out of the public sphere. But exclusion is neither a possible nor a practical option. First of all, there are simply many evangelicals. They make up 25–33 percent of the Protestant population (Mitchell and Tilley 2004). Second, evangelical symbolism, assumptions, and mores still appeal even to some nonchurchgoing Protestants, informing the "Protestant ideology" (Wright 1973; see also Todd 1987) and forming the core of Protestant ethnic identity (Bruce 1986). Third, evangelicals are—by definition—committed to activism. This predisposes them to "do something." This is significant in and of itself, because of the relative lack of activism within the wider Protestant community. Cochrane and Dunn (2002) have contrasted Catholics' vibrant community-based activism with that of Protestants. Although Farrington (2001) and Cochrane and Dunn and point to an increase in loyalist working-class activism since 1972, the gap between Catholic and Protestant community activism persists. Many Protestants are reluctant to identify with loyalist activism, especially if it is associated with former paramilitaries (Langhammer 2004; Gormally 2001; McEvoy 2001). Evangelicals, however, have been increasingly prominent in promoting church-based activism and forming special-interest organizations. The churches remain Protestants' largest community-based structures (See Morrow et al. 1991; Bacon 2003, 1998).[10] Transforming the conflict in Northern Ireland requires taking into consideration evangelicals from Paisley to ECONI.

The Content of Northern Irish Evangelical Beliefs

The emergence of evangelicals who self-consciously and publicly distanced themselves from Paisley was significant. ECONI saw part of its task as critiquing the Calvinist-informed theology that had been dominant throughout Northern Ireland's history (Thomson 2002, 1996). Although Northern Irish evangelicalism is more diverse than Paisley and ECONI (Mitchel 2003; Jordan 2001; Brewer and Higgins 1998; F. Porter 2002), the most active evangelicals clustered around these "traditional" and "mediating" poles (see chapter 1). Traditional evangelicals have continued to articulate positions that echo the Calvinist-influenced beliefs outlined above. The most systematic presentation of the themes endorsed by mediating evangelicals has come through the work of ECONI.

ECONI has borrowed heavily from the Anabaptist tradition (see Yoder 1994; Hauerwas and Willimon 1995,1989; Hauerwas 1995, 1983; C. Carter 2001).[11] The classic Anabaptist position on the relationship between church and state is that they should be totally separate. Any sort of close relationship with the state is seen to compromise the church's ability to play its proper role as an alternative sociopolitical order. Being an alternative sociopolitical order involves rejecting political power. If the church does not reject power, its interests become bound up with the state, leading to "idolatry"—the worship of the state as well as (or instead of) God. ECONI drew on these concepts to argue that Northern Ireland's covenantal Calvinism had made evangelicals into idol-worshippers. ECONI claimed that some evangelicals had placed their allegiance to Ulster ahead of their allegiance to Christ. ECONI viewed attempts to ensure that "right religion" has a privileged relationship with sociopolitical power as misguided. This would not bring about God's blessing; rather, it would inhibit the church from doing what it was intended to do.

Insistence on the separation of church and state also shapes the Anabaptist attitude about religious and cultural pluralism. The ideal Anabaptist world is one "in which no one is forced either by the government or by societal expectations to be Christians" (Hauerwas 1995:73). This allows the church to maintain its distinct identity while at the same time leaving space for the toleration of other forms of Christianity, other religions, and other cultures.[12] In the Northern Ireland context, this might mean that a sharp contrast between Protestantism and Catholicism remains. However, ECONI has made it a point to engage with nonevangelical Protestants and Catholics. They do so from a position wherein their evangelical identity is not threatened by religious or cultural pluralism.

Finally, the Anabaptist emphasis on suffering and servanthood provides no justification for violence. Pacifism is a sacred principle for classic Anabaptists. It is based on the nonviolent example of Jesus in the Gospels, specifically his rejection of political power, his acceptance of his role as servant, and his willingness to suffer. Violence is considered illegitimate, even if it is used to achieve noble ends. The use of nonviolent protest is considered an effective technique, as practiced by Martin Luther King Jr. or Gandhi. But effectiveness is not the point. The Anabaptist tradition recognizes that nonviolence, more often than not, seems to fail. After all, Jesus' death on the cross at first seemed a failure. What is essential is a commitment to patience, not effectiveness. The church must be willing to suffer indefinitely, confident that God will maintain its survival. ECONI, as well, has articulated pacifist positions.[13] ECONI's pacifist strand has led to an organizational focus on issues of peacemaking, such as forgiveness and reconciliation. These beliefs provide mediating evangelicals with a theological justification for cross-community activism and peace-building projects.

Evangelicals and Religious Networks in Contemporary Northern Ireland

Traditional and mediating evangelicals have translated their convictions and concerns into well-organized networks of special-interest groups and congregations, which are explored in detail in chapters 4 and 5. However, evangelical activism has not been adequately accounted for and analyzed in most research on religion in contemporary Northern Ireland. Rather, it has focused on religious actors who are involved in cross-community groups or activities. For example, Power (2007) documents the contributions to reconciliation of national-level ecumenical dialogue, local churches, ecumenical communities, and peace and reconciliation education initiatives. She argues that the ecumenical movement has both reinforced and contributed to the development of a community-relations-based approach to reconciliation. In fact, Appleby (2000) has argued that these religious actors have been amongst the most effective in the world in contributing to conflict transformation. This may come as a surprise to observers of the Northern Ireland conflict! However, Appleby's analysis is based on the activities of Catholic and Protestant religious actors and focuses largely on ecumenical activity.[14] He describes the wider peace movement in Northern Ireland as a network of relationships between religious and cultural organizations. For him, Northern Irish society is "saturated" with religious peacebuilders who are willing to cooperate with secular peacebuilders. This has made it a particularly promising site for the transformation of conflict:

> Compared to most conflict settings around the world, Northern Ireland is saturated with religious and cultural practitioners of conflict transformation. Further, because the peace advocates operated at

several levels of religion and society and persisted through decades of continuous activity, they became part of the institutional and social landscape. At the highest official levels of the Catholic, Presbyterian, Methodist, Anglican, and other churches...religious leaders condemned sectarian violence, criticized their belligerent coreligionists, encouraged peacemaking efforts, entered into ecumenical dialogue with one another, and sponsored joint social, economic, and educational initiatives designed to foster cross-communal cooperation and build trust among erstwhile antagonists. At the level of middle-management—parish- and congregation-based religious leadership—a few extraordinary priests and ministers served as intermediaries between Catholic or Protestant paramilitaries and church and government officials seeking solutions to the conflict. Parachurch reconciliation groups such as Corrymeela, Cornerstone, and ECONI operated in yet another sector of society, among educators, politicians, professionals, and working-class victims of the Troubles. Other religious and cultural actors participated in community organizing groups. (Appleby 2000:236–237)

Although Appleby mentions the activities of evangelicals (in particular, ECONI and Rev. Ken Newell, the minister of Fitzroy Presbyterian in Belfast), he does not emphasize that evangelicalism has been more significant for the Protestant/unionist community than Catholicism has been for the Catholic/nationalist community. This may, in part, explain his optimistic analysis of the role of religious actors in conflict transformation. Appleby hints at this when he discusses the different communal reactions to ecumenical activities, as experienced by the Presbyterian Rev. John Dunlop and Cardinal Cathal Daly:

Ecumenically minded Protestant clerics who supported Dunlop had to contend with pressure from members of the Orange Order who paid them visits whenever they seemed to be "going soft" on the conflict. Dunlop himself faced critics from within the Presbyterian General Assembly as well as the troublesome "ultras" clustered about Paisley. Cardinal Daly confronted the reality that a large segment of the nationalist population listened only selectively to the words of the Catholic Church or had abandoned it altogether. (2000:184)

Appleby's examples show that Dunlop and his supporters faced *active religiously motivated* opposition, whilst Daly's opposition was muted, indifferent or not particularly religiously motivated. Appleby does not sufficiently emphasize that the religious dimension has been more important to the Protestant community. As such, understanding the role of *ecumenical* Christian peacemaking in Northern Ireland does not give us an adequate picture of religion's role and potential contribution to conflict transformation (see also Tombs and Liechty 2006). This is not to doubt the religious commitment, sincerity, or even effectiveness of ecumenical activists, but it is to argue that their religiously based discourses and activities will have a limited appeal as uniquely *religious* contributions to conflict transformation. The work of ecumenical activists does not appeal to many evangelicals. Because of evangelicalism's historical significance, it is evangelical networks that have greater potential to speak to them with a prophetic voice. If evangelicals (as opposed to other religious or secular actors) offer a reworking of beliefs or justifications for peacemaking, they may be more likely to receive a hearing. No previous research has provided a systematic account of evangelical activism since the Belfast Agreement. This leaves our understanding of religion's contribution to conflict transformation incomplete.

Evangelical Activism in Comparative Perspective

In neither the USA nor Canada has evangelicalism been bound up in a conflict with a religious dimension.[15] But each provides useful points of comparison because they are both societies in which evangelicalism once held a privileged position, and evangelicals there have had to adjust to the breakdown of their relationship with social and political power. In addition, the USA and Northern Ireland are amongst the societies with the highest proportions of evangelicals in the world.[16] According to the 1997 Angus Reid World Survey of "True Believers,"[17] 28 percent of American Protestants fall into this category (Noll 2001a:41; Noll 1998). Northern Ireland was not part of the Angus Reid survey,[18] but most estimates of the evangelical population put it at 25–33 percent of the Protestant community (Bruce 1986; Boal et al. 1997; Mitchell and Tilley 2004). In Canada, only 8 percent of Protestants are ranked as "true believers." But Canada is still quite evangelical by international standards. Besides the USA, Northern Ireland, and South Africa, the only societies with a higher percentage of Protestant "true believers" than Canada are Brazil, the Philippines, and South Korea with 10 percent each.[19] As in Northern Ireland, this provides evangelicals with a solid base of activists.

The remainder of the chapter provides a historical overview of evangelicalism's relationship with power in the USA and Canada. By demonstrating how evangelicalism played a priestly role in each society, it establishes the relevance of the comparison to Northern Ireland.

Then, it analyses how evangelicals in the USA and Canada have adjusted their activism in light of their loss of privilege. Finally, it relates the comparative analysis to Northern Ireland.

Historical Overview

USA

Protestants in the American colonies were divided amongst various forms of Christianity, but covenantal Calvinist traditions were probably the most influential at the time when evangelical revivalism met American shores.[20] The Massachusetts Bay colony, founded by English Puritans, was the prototype. The evangelists preached messages of liberty and freedom for the individual believer. These messages tended to be transferred to the political arena, uniting the colonists against what was perceived as an oppressive British government. Some have argued that the Great Awakening of the 1740s was the impetus to the American Revolution (Heimert 1966). At the very least, Noll, Hatch, and Marsden (1989) argue that during this time evangelicalism altered and retained certain aspects of Calvinist political theory. American evangelicals, on the one hand, retained the sense that Americans were a chosen people but, on the other hand, abandoned the Calvinist sense of the "inability (even of the Christian) to serve God perfectly" (Noll, Hatch, and Marsden 1989:42). This meant that they did not reflect theologically on the ideologies that informed the Revolution but adopted those ideologies indiscriminately, equating "loyalty to the new nation and loyalty to Christ" (Noll, Hatch, and Marsden 1989:63). Now it was republican individualism that was considered truly Christian and marked Americans as a chosen people. This brand of evangelicalism came to serve as a unifying creed after the Revolution (Marsden 1980; Watson 1997; Monsma and Soper 1997). Evangelicalism played a priestly role, its prominence evidenced by significant conversions in the west and south; evangelical influence in mass communications, popular thought, education, and the values of political parties; and the respect bestowed upon evangelical theologians. As with the Revolutionary War, a number of links can be traced between the activism inspired by the Second Great Awakening and the tensions that resulted in the Civil War (1861–1865).

Evangelicalism's priestly, unifying role was challenged during the Civil War. Evangelicals, like their countrymen, divided along the Mason-Dixon Line. They used religious arguments to justify their political causes.[21] It might have been expected that the North's victory would be interpreted as God's vindication of their cause, and, to some extent, it was. However, the war had the paradoxical effect of producing "a nation in which the power of religion declined":

> [A]s the conflict raged, believers in both the North and South were tempted to make the success of their military efforts the object of their most basic religious concern. To the extent that this took place, a most unfavourable precedent was established. By making such strong commitments to the righteousness of their own side and by regarding the enemy in such deeply religious terms, believers set the stage for other consuming national interests to exert a shaping influence on the churches...To reapply an old saying first used in the early history of Christianity, the believers who married the spirit of the Civil War age found themselves widowed in the age that followed. (Noll 2001b:323)

During that "widowing" process, evangelicalism's relationship with social and political power-holders began to break down. Evangelical America had assumed that progress was linked to biblical Christianity and that "faith, science, the Bible, morality and civilization" were interrelated (Marsden 1980:15). But higher criticism, Darwin's theory of evolution, and World War I challenged this view of the world. Evangelicals were divided about how best to cope with these challenges. Some advocated separation from society, some argued that evangelism was the best strategy, others argued that Christian civilization could be preserved, and still others sought to "transform" culture (Marsden 1980:136). The theological debate between modernists and evangelicals about higher criticism took on new dimensions when the supposed effects of higher criticism ("German Barbarism," Marsden 1980:149) were linked to the theory of evolution and the war. This debate united a significant number of evangelicals, who became known as fundamentalists.

The fundamentalist-modernist debate culminated in the Scopes trial. In 1925, Tennessee passed a law prohibiting the teaching of Darwinism in public schools, which biology teacher John Scopes challenged in the courts. William Jennings Bryan, hitherto a nationally respected evangelical figure, volunteered to defend the fundamentalist position.[22] The trial, staged in rural Dayton, Tennessee, became "the symbolic last stand of nineteenth-century America against the twentieth century" (Marsden 1980:185). The twentieth century won. The severing of the links among evangelicalism, cultural prominence, and sociopolitical power seemed complete. Fundamentalists retreated to lick their wounds, increasingly advocating separatism and pietism. They went about quietly building separate institutions, such as Bible schools, and were relatively absent from public life. However, the loss of their privileged position did not mean that American evangelicals would never again impact society and politics. American social and political structures (examined in more detail below) provided a framework within which well-organized groups, such as evangelicals, could agitate effectively—and prophetically—in the public sphere.

Canada

In colonial Canada, evangelicalism was strongest in the Maritimes, where a "majority of the residents in the 1760s were immigrants from [evangelical and Puritanical] New England" (Noll 2001b:126–127). However, Henry Alline, the most influential revivalist preacher in the Maritimes, preached an apolitical, anti-Calvinistic Gospel and settlers versed in his "New Light" message were loathe to draw the links between religion and revolution that had been drawn in the American colonies. Maritime evangelicalism became pietistic, not Calvinistic.[23]

In other regions of the Canadian colonies, evangelicalism encountered and adapted to the Christian traditions already on the ground. In Upper Canada (modern Ontario), Anglicanism and its regard for tradition and hierarchy meant that evangelicals there did not readily articulate a discourse linking freedom and individuality, as they had done in America. Colonial Canada also had proportionately a far greater number of Catholics than colonial America. Quebec was almost entirely Catholic[24] and there were significant numbers of Catholics in Acadian New Brunswick.[25] Noll argues that the need for Catholics and Protestants to coexist led to a tradition of accommodation and tolerance. The balance of sociopolitical power between Catholics and Protestants was not equal, but it was "equal enough" to ensure that tolerance rather than domination by one group or the other was the option that emerged. In Ontario and in western Canada, "divergent strands of Protestant experience at the start of the nineteenth century eventually came to assist each other in the creation of an *evangelical* Protestant civilization" (Noll 2001b:266, emphasis mine; see also Gavreau 1991 and Choquette 2004). This concept of civilization was rooted in the idea that the churches should play a public role, working closely with the state to promote order and stability. It was exhibited in a tendency toward consensus building and cooperation. In this way, evangelicalism played a priestly role. However, Canadian evangelicals generally refrained from a "chosen people" discourse (Noll 2001b; Bebbington 1997; Simpson and MacLeod 1985; Handy 1982; Dekar 1982).

The Canadian concept of an evangelical Protestant civilization was also impacted by events in America. Canadians had already rejected the ideals of the American Revolution. The War of 1812 focused and consolidated these trends.[26] The American Civil War confirmed for Canadians that they had been right to "avoid taking the American path" (Noll 2001b:249). The formation of modern Canada in 1867 reflected a wariness of the expanding nation to the south, as well as a desire to maintain a Loyalist tradition that was suspicious of republicanism and unfettered democracy. This tradition had a high regard for slow, evolutionary change and compromise.[27] Canadian evangelicalism reflected and shaped these values. World War I and other changes at the turn of the century such as "the immigration of non-Protestants, the growth of industrial giantism, and new ideas about science and the Bible" presented a challenge to Canadian evangelicals (Noll 2001b:275). But rather than experiencing a process of "widowing," Canadian evangelicals' tendency to act as mediators meant that they received these upheavals with "greater calm" than American evangelicals (Noll 2001b:275). The more significant challenge to Canadian evangelicalism's priestly role has been widespread secularization, especially from the 1950s onward (Bibby 1990). Secularization has caused a degree of consternation amongst evangelicals, especially when it seemed as if Canadian evangelicals' tendency toward pietism might gain predominance over their tendency toward mediating activism (Stackhouse 1997; Bibby 1993; Grenville 1997; Stiller 1996). Activist-oriented evangelicals argued that the pietist strand would trump the mediating strand if evangelicals did not become more aware. However, Canadian evangelicals did not disappear from society and politics altogether. Rather, they devised strategies of mediation that embraced cultural pluralism and worked well within centralized Canadian political structures. This has meant that at particular points in time they have raised a discrete—but prophetic—voice in the public sphere.

Parallels with Northern Ireland

Several important parallels can be drawn among the historical roles of evangelicalism in Northern Ireland, the USA, and Canada. In Northern Ireland and the USA, Calvinist-inspired conceptions of the covenant and the chosen people were prominent. In the USA, this provided evangelicals with a theological justification for republican individualism. It also led them to develop an uncritical stance toward the state and its power. In Northern Ireland, these concepts provided evangelicals with a theological justification for maintaining a boundary with Catholics, and for their privileged sociopolitical position vis-à-vis Catholics. This led evangelicals to develop an uncritical stance toward the Stormont government and its power. To the extent that these Calvinist concepts continue to inform evangelical belief in the USA and Northern Ireland, parallels between evangelicalism in the two contexts may continue to be drawn.

In Canada, on the other hand, the imbalance of power between Catholics and Protestants was not as great as it was in Northern Ireland. Though there were tensions and boundaries (particularly with Catholic Quebec), Catholics and Protestants in Canada managed to negotiate a more peaceful, pluralist political order. The prominence of anti-Calvinist, Anabaptist, and pietist theologies in Canada also worked against the development of covenantal and chosen people discourses. To the extent that the balance of power between Catholics and Protestants in Northern Ireland is evening out, they may begin to negotiate a more peaceful, pluralist political order. As in Canada, the emergence of anti-Calvinist evangelical theology in Northern Ireland may contribute to this process.

Contemporary Evangelical Activism
in the USA and Canada

USA

By the 1940s, evangelicals had regrouped and were ready to reenter public life. Key individuals formed organizations such as the National Association of Evangelicals (NAE) (which repudiated fundamentalist positions) and Youth for Christ (led by Billy Graham) (Wuthnow 1988; C. Wilcox 1992; Wilcox and Rozell 2000). Evangelicals were active in the Christian Anti-Communism Crusade of the 1950s and were relatively united in their support of Republican Barry Goldwater's presidential candidacy in 1964 (C. Wilcox 1992; Wilcox and Rozell 2000).

Sweeping structural changes in American society during the 1960s contributed to a proliferation of special-interest organizations, both sacred and secular (Wuthnow 1988).[28] Wuthnow notes that these changes contributed to a shift to the left amongst religious denominations and congregations—even amongst some evangelicals. For instance, the evangelical Sojourners community formed during this time and some African American evangelicals began to criticize the

Vietnam War. At the same time, other evangelicals mobilized with conservative goals in mind, forming groups such as the Creation Research Society and Americans for God. This led to a polarized realignment within evangelicalism. The election of a self-proclaimed "born again" president in 1976, Democrat Jimmy Carter, brought a renewed visibility to evangelicalism in the public sphere. Many evangelicals, however, soon felt let down by the performance of the Carter administration. By the late 1970s, what has been called the "New Christian Right" began to mobilize (Bruce 1988; Liebman and Wuthnow 1983).

The New Christian Right used special-interest organizations to great effect. This movement was centered around the personalities of religious broadcasters such as Pat Robertson, Jerry Falwell, Jim Bakker, and Jimmy Swaggart; and organizations such as the Moral Majority, Christian Voice, and the Religious Roundtable, all founded in 1979 (Wuthnow 1988; Hunter 1987; Guth et al. 1996). They claimed that the USA had been founded by Christians and that its government should have a Christian ethos. They argued that America was becoming increasingly immoral and must repent or face God's judgment. They focused on issues such as abortion, school prayer, pornography, and homosexuality (Wuthnow 1998; Guth 1996). This group, with its high visibility and strident claims, soon came to be regarded as dominant within evangelicalism.

But by the end of the 1980s, the Moral Majority and other prominent national organizations had disbanded. Robertson turned his energies toward forming the Christian Coalition, which came to be regarded as the national voice of the movement. Under the leadership of Ralph Reed (1989–1997), the Christian Coalition concentrated on grassroots organizing and formed a variety of coalitions with secular conservative politicians, conservative Catholics, Mormons, and Orthodox Jews (Watson 1997; Wallis and Bruce 1986; Moen 1992). The Christian Coalition's grassroots organization has been particularly effective and has enabled the religious right to become perhaps the most important grouping within the Republican Party at both the local and national levels (Jelen 2002; C. Wilcox 1992).[29] Most recently, the Christian Right has been credited with securing the election of President George W. Bush, largely because of his "born again" faith and his conservative positions on abortion and gay marriage (Trends 2005).

At the same time, the rhetoric of the Christian Right has shifted. Although they have not abandoned talk of a "Christian America" altogether, they have begun to demand a "place at the table" in a

plural, civil society (Jelen 2002; Watson 1997). Watson (1997) argues
that the New Christian Right has adopted the language of other multicultural groups, such as women and ethnic minorities, who argue
for nondiscrimination. Because the New Christian Right continues
to argue for a Christian America *and* for a place at the table, they have
often been accused of hypocrisy or of trying to make the USA a
Christian nation by stealth. Watson has another explanation:

> [T]hey want both. They want "their place at the table," *and* they want
> everyone at the table to agree with them. They want a Christian nation
> *and* religious freedom. As contradictory as it may sound, they want to
> have their cake *and* to eat it too. (1997:175)

American evangelicals' activism strategies have been shaped also by
sociostructural conditions. For instance, Wallis and Bruce identified
four structures that facilitate evangelical activism (1986:306): government, administration, mass media, and party system. First, in the
USA, government is federal and diffuse, which allows smaller blocks
such as evangelicals to gain a relatively high degree of control at local
levels, if not at national levels. Second, administrative offices are filled
by election, and well-organized minorities are able to "become an
electoral force" in such situations, where multiple elections interact
with widespread voter apathy (Wallis and Bruce 1986:309). Third,
the media is diffuse and open, which allows well-organized blocks of
evangelicals to buy plentiful airtime and print space. And finally, the
party system in the USA is weak, which allows for lobby groups
and political action committees to have a disproportionate influence
on party politics. This point is underlined by the way in which evangelicals of the right have been able to carve out a niche for themselves
within the Republican Party (Jelen 2000; Kellstedt et al. 1994; Wald
2003; Rozell and Wilcox 1997; M. Wilcox 2002).

In sum, American evangelicals have responded to the opportunity
to play a prophetic role by adjusting both their social activism strategies and their theologies. Evangelical activism is dominated by a vocal
religious right that focuses on moral issues and is closely aligned with
the Republican Party. Evangelicals have pragmatically cooperated
with Catholics, Jews, and secular allies on particular (usually "moral")
issues. They demand their "place at the table" in a competitive, plural
civil society. However, much evangelical rhetoric continues to be
laced with "Christian America" themes that reflect historic, Calvinist
conceptions that mark Americans out as a chosen people. This means
that many of their erstwhile allies are not convinced that evangelicals

do not *really* want a Christian America. This contributes to a "culture wars" atmosphere in which cooperation may be more difficult (Hunter 1991). However, it is not clear whether a majority of American evangelicals agree with the New Christian Right agenda; indeed, many actively disapprove (Smith 1998; Hunter 1991, 1987; Buell and Sigelman 1987, 1985; Sigelman, Wilcox, and Buell 1987; C. Wilcox 1992, 1987; Jelen 1991). Some of these evangelicals are not engaged in social activism; they have opted to withdraw. Others agree with parts of the New Christian Right agenda but dislike their tone and the strident way they make their case in the public sphere. A minority of evangelicals, such as those associated with the Sojourners community or Tony Campolo, could be classified as part of the religious left (Gasaway 2004; M. Lindsay 2004).[30] Finally, evangelical activism is facilitated by social structures that encourage interest-group activity and active participation in grassroots party politics. The strategies and rhetoric currently emerging amongst traditional evangelicals in Northern Ireland have parallels with those used by the New Christian Right in the USA.

Canada

Unlike in the USA, secularization in Canada has been a major story in religion during the twentieth century and has contributed greatly to evangelicalism's loss of its privileged position (Bibby 1990). However, Canadian evangelicals seem to have been immune from the secularization process, partly due to their immersion in the evangelical subculture, in which public religious practice is heavily encouraged (Reimer 2003).[31] There are proportionately no fewer evangelicals in Canada than before the attendance drop off, and they are as religiously committed as ever.

Throughout the twentieth century, evangelicals' religious commitment has translated into periodic bursts of mediating activism. Noll (2001b:276–78; 2001a:237–240) cites a number of examples, arguing that evangelical-influenced political projects, such as the "rightist Social Credit of fundamentalist preacher William Aberhart and the leftist Cooperative Commonwealth Federation (CCF) led by ordained Baptist minister Tommy Douglas" have had a tendency to last longer and "exert broader effect than comparable efforts in the United States" (2001a:251). Rawlyk argues that the New Democratic Party (NDP) has continued to be influenced by its evangelical and "Social Gospel heritage" (1990:269). Douglas led the NDP from 1961 to 1971.[32]

But evangelicals' religious commitment has not always translated into mediating activism. Stackhouse argues that evangelicalism has been inward-looking and pietistic, "with no discernible influence upon Canadian public life" (1997:67; Bibby 1993). Grenville (1997) observes that 3.2 million Canadians have had evangelical religious experiences, although only 1.1 million of them have had contact with organizations such as the Evangelical Fellowship of Canada (EFC) or the Inter-Varsity Christian Fellowship (IVCF); this observation underlines the fact that most Canadian evangelicals are more pietistic than activist. Even the evangelical/charismatic revivals associated with the Toronto Blessing seem to have led to more personal, privatized religious revitalization rather than social activism (Poloma 2003).

On the other hand, there are signs that Canadian evangelicals have become more committed to activism as they perceive the erosion of Canada's Christian heritage. From the 1960s onward, evangelicals began to form more organizations, such as the EFC, Citizens for Public Justice, a Canadian office of the American group Focus on the Family, and the Centre for the Renewal of Public Policy (Stackhouse 1997).[33] The EFC was formed in 1964 but experienced dramatic membership and financial growth during the 1980s (Stackhouse 1995). Stackhouse (1995) notes that the EFC has been the largest and most visible evangelical political organization in Canada, and its mode of engagement has been largely what Noll would call mediating.[34]

Canadian evangelicals are troubled because the society around them no longer promotes "Christian" values (Stiller 1991a, 1991b, 1996; Cannell 1996, Stackhouse 1990). And like American evangelicals, they have started to argue for their "place at the table" (Stiller 1997). However, Canadian evangelicals have not aligned themselves with a political party to the same extent as American evangelicals. Canadian evangelicals are more likely to support the conservative Reform party than nonevangelicals, but American evangelicals are twice as likely to support Republicans as Canadian evangelicals are Reform (Reimer 2003:127; Noll 1998; Guth and Fraser 2001).[35] Americans are also "twice as likely as Canadians to consider religion very important to their political thinking, and American evangelicals follow this tendency as well. Americans and American evangelicals also are slightly more likely to think traditional values should inform politics and more likely to hear political messages from the pulpit" (Reimer 2003:128).

Canadian evangelicals' rhetoric and their theological justifications for their activism are very different from those found in the USA. This is reflected in the writings of Brian Stiller, former president of the EFC. Arguing that "cultural pluralism" is an opportunity for evangelicals to impact the Canadian public sphere, he writes: "For Canadian Christians, especially those from a European or American background, [cultural pluralism]...is an opportunity to unlink Christian faith from position and history as a means of imposing Christ's ministry on society" (Stiller 1996:115). Stiller even goes so far as to say that "cultural pluralism is...a basic Christian affirmation that we—as God does—are to give space and allowance for people to think, believe, act and hope with different assumptions than those offered by God's revelation in Christ and the Scriptures" (1996:109).

Canadian evangelicals' activism strategies are shaped also by socio-structural conditions. For instance, features of the political system make party alignment less profitable for interest groups than it does in the USA. As Hoover notes, the aggressive special-interest-group tactics so popular in the USA would not work in Canada, "because the concentration of power in the hands of government elites limits the influence of such groups. The aggressive populist 'outsider' tactics of interest groups are often met with indifference by political leaders, whereas the discrete and respectful development of government allies may be necessary for 'inside' influence" (in Reimer 2003:132). That means their political strategies include cultivating discrete relationships with government allies across the political spectrum. It also encourages them to cultivate working relationships with nonevangelical groups. Finally, tighter broadcasting controls than in the USA limit the amount of publicity that may be secured by Canadian evangelicals.

In sum, Canadian evangelicals also have responded to the opportunity to play a prophetic role by adjusting their social activism strategies and their theologies. Their activism is marked by a mediating approach that seeks to cultivate discrete relationships with politicians and other civil society groups. Canadian evangelicals conceive their place at the table not so much as an unfortunate relegation, but as a base from which evangelicals can impact society without calling for a "Christian Canada." In this, contemporary Canadian evangelicals are in continuity with Canada's anti-Calvinist evangelical strand. They are pushing that strand to its limits, with Stiller even going so far as to say that cultural pluralism "is a basic Christian affirmation" (1996:109). This style is well suited to Canadian sociopolitical structures, which discourage conflictual interest-group activity and encourage groups to build discrete relationships with government and amongst themselves. The diffuse character of Canadian evangelical activism makes it difficult to say with any degree of certainty just how much evangelicals are impacting sociopolitical processes. A continuing strong pietist strand also works against Canadian evangelicals having a far-reaching impact. However, Noll (2001b; 2001a), Dekar (1982), and Stackhouse (1997) have argued that their subtleness means that Canadian evangelicals have had a greater long-term impact than American evangelicals. They cite particular causes that Canadian evangelicals have championed and subsequently have been diffused widely in the rest of society. The strategies and rhetoric currently emerging amongst mediating evangelicals in Northern Ireland have clear parallels with those used by Canadian evangelicals.

Contemporary Northern Irish Evangelical Activism in Comparative Perspective

These comparative perspectives provide insights into possible future directions for Northern Irish evangelical activism (table 3.1). For example, in the USA, as evangelicals lost their privileged position the dominant form of activism that emerged focused on moral issues, interest-group activity, and cultivating links with the Republican Party. Sociostructural conditions such as diffuse government structures and an open media facilitated this activism. American evangelicals still claimed they wanted a Christian America, but they were willing to settle for a place at the table. There are parallels here that are relevant to the situation emerging in Northern Ireland. For instance, given Northern Ireland's evangelical history, influenced as it is by "chosen people" Calvinism, activism may continue to move in an American direction. This already seems to be the case amongst traditional evangelicals, who say they fight "for God and Ulster," whilst simultaneously using the language of pluralism and nondiscrimination. The Assembly also gives government structures a federal and diffuse character, as it does in the USA. Now that the Assembly is no longer suspended and there is a greater possibility that more powers will be devolved to local government, American Christian Right strategies such as aggressively targeting local elections might be effective in Northern Ireland. The DUP's historical association with evangelicalism presents another opportunity for evangelicals to try to gain influence. However, evangelicals in Northern Ireland must negotiate tighter media controls than those in the USA; how successfully they negotiate may impact their ability to get their message out.

There is potential for traditional evangelicals to diverge from American evangelicals in their willingness to cooperate with secular

Table 3.1 Evangelical Activism in the USA, Canada, and Northern Ireland

	USA	Canada	Northern Irish Traditional	Northern Irish Mediating
Forms of Activism	Focus on moral issues; Republican Party; use interest-group tactics	Periodic single issue focus; seek to "mediate" within civil society and between government and civil society	Focus on moral issues; DUP; beginning to use interest-group tactics	Focus on peacebuilding and social justice issues; seek to "mediate" within civil society and between government and civil society
Theological Emphases	Covenantal Calvinism ("Christian America") coupled with pragmatic acceptance of pluralism	Anti-Calvinism and pietism; enthusiastic about pluralism	Covenantal Calvinism ("For God and Ulster") coupled with pragmatic acceptance of pluralism	Anti-Calvinism and Anabaptism; enthusiastic about pluralism
Structures	Federal and diffuse government structures and open media encourage interest-group activity	Centralized government structures and regulated media encourage discrete, behind-the-scenes activism	Possible federal-style devolution (if the Assembly and local governments gain more power) could encourage interest-group activity	Centralized government structures (during direct rule and currently) and regulated media could encourage discrete, behind-the-scenes activism; Government preferences for cross-community activity favors "mediators"

or, more obviously, Catholic allies. Wallis and Bruce have argued that this is *the* critical difference in Northern Ireland (1986:317). However, as sociostructural conditions have changed and the power disparity between Catholics and Protestants has narrowed, the impetus for a negotiated, peaceful coexistence (as in colonial Canada) has increased. In addition, Wallis and Bruce wrote before the Belfast Agreement and before many of the British government's civil society reforms had taken effect. The British government's civil society approach may favor the development of mediating evangelicalism, which rejects the propriety of evangelicalism's former privileged position and favors cross-community initiatives. Indeed, mediating evangelicals are cooperating with Catholics and other groups to a much greater extent than would be expected from Wallis and Bruce's analysis. They are receiving government support for doing this. They also developed close, discrete relationships with government officials under direct rule. These relationships are similar to those developed by Canadian evangelicals within Canada's more centralized political structures. Finally, Wallis and Bruce wrote before mediating evangelicals became noticeable in the public sphere. Mediating evangelicals have undertaken an ambitious project to change evangelical theology and identity, deliberately rejecting "For God and Ulster" claims. Like Canadian evangelicals, they critique Calvinism and are enthusiastic about pluralism. Mediating evangelicals have developed strong links with evangelicals in the Republic of Ireland, a small minority that is currently seeking to develop a mediating role.[36] As evangelicals attempt to take on this role in the Republic of Ireland, they may come to have a greater influence on some evangelicals in Northern Ireland. These trends suggest that Northern Irish evangelicalism could also move in a mediating direction similar to evangelicalism in the Canadian context.

Conclusions

Evangelicals' prominent place in the Protestant community means that their activism has potentially far-reaching effects. This comparative framework does not allow us to make predictions about whether Northern Irish evangelicals will follow an American or a Canadian path. It does suggest that the conditions are present (both in sociopolitical structures and in content of belief) for Northern Irish evangelicals to move in either (or both) directions. These moves have implications for evangelicalism's potential to contribute to conflict transformation.

To the extent that traditional evangelicals follow an American trajectory and uphold Calvinist concepts, they may continue to be the guardians of Protestant ethnonational identity. However, they may adapt to the language of pluralism and nondiscrimination, an adaptation that is obvious in the discourses of the DUP (Ganiel and Dixon 2008; Ganiel 2007; Rankin and Ganiel 2007). This adaptation may not signal a quick and enthusiastic embrace of the new order. But it is not an outright rejection of it either, which indicates that traditional evangelicals may not be the obstructionist enemies of the Belfast Agreement that they are often assumed to be. Rather than simply being cranks, they may be raising valid criticisms of the Belfast Agreement and how it has been implemented.[37] They are changing, albeit slowly and painfully, and their ability to adapt (whilst watching the process with a critical eye) may reinforce similar trends in the wider Protestant community.

As for mediating evangelicals, their potential contributions to conflict transformation seem more obvious. After all, their goals are to "fix" aspects of evangelicalism that they believe contributed to conflict. They want to change evangelical identity and to convince evangelicals (and the wider Protestant community) to become involved in building a new, pluralist order. Their reworking of evangelical theology taps into the deep cultural respect for the Bible that is held in the wider Protestant community. They have embraced the new order and can provide evangelicals and other Protestants with culturally and theologically relevant justifications for doing so. Canadian evangelicals have engaged in a similar process. This has the potential to contribute to the transformation of Protestant identity into one that is secure in itself and enthusiastic about pluralism. However, it is not clear to what extent mediating evangelicalism has penetrated the grass roots. As such, the mediating form of evangelical activism also may be slow in producing its desired results.

Chapter 4

Evangelical Congregations and Identity Change

In Northern Ireland, evangelicals within congregations have been forced to respond to widespread social and political changes, and this has contributed to changes in the content of evangelical identities. Focusing on the way people frame their identities avoids too immediate analysis of reactions to specific policies and gets at underlying trends. As Todd (2005) has argued, it allows for analysis of how changing identities impact wider social and political processes, including generating crises as well as opportunities for acceptance of the postconflict order. Todd's typology of identity change, outlined in chapter 2, provides an underlying framework for this chapter.

The chapter proceeds by outlining the theoretical and practical considerations in the selection of the congregations under study. It then analyses the links between identity, change, and activism within each congregation. It concludes with an analysis of how identity change is impacting wider social and political processes, including the transformation of conflict.

Selection of the Congregations

Theoretical Considerations

In order to capture the full diversity of Northern Irish evangelicalism, I wanted to locate congregations in which I would find evangelicals holding a range of beliefs on the traditional-mediating spectrum. This is important, because much prior research on Northern Irish evangelicalism has focused on Paisleyism and failed to capture either the diversity within evangelicalism or the changes that have been taking place over the last few decades. Mitchel's (2003) work has gone some way toward mapping changes in evangelical identity, but it is limited to an analysis of the documents and articulations of elite spokesmen. This likely overemphasizes the importance of theological reflection in the process of change and raises questions about identity changes among nonelites.

Grounding the research in congregations allows for the consideration of factors that may contribute to the process of identity change. Congregational factors can be held constant so that other factors that contribute to changes within individuals may be isolated. Ammerman et al. (1998) analyze congregational factors in terms of four "frames." Each frame may enable and/or constrain the ability of individuals within the congregations to act and to change.[1] The ecological frame is the environment in which the congregation, including other social and religious organizations and political institutions, is located. This environment may vary within states or political jurisdictions. For instance, the environment surrounding rural or urban congregations in Northern Ireland may be different, so the process of change that takes place within these congregations may also be different. The cultural frame considers the congregation's distinct identity, which people within it have constructed together over time. It "includes all the things a group does together—its rituals, its ways of training newcomers, its work, . . . its play," its artifacts (such as buildings and newsletters), and the stories it tells about itself (Ammerman et al. 1998:15). Congregational cultures provide people with a way of thinking about the world and embodying their faith, acting as a powerful facilitator or inhibitor of change. For instance, if individuals begin to change in ways that put them at odds with the congregational culture, they may either decide to leave or attempt to stop the processes of change within themselves. On the other hand, a congregational culture that conceives of change as compatible with faith may encourage the process of identity change. The resources frame consists of the potential "capital" that members of the congregation may draw on when experiencing processes of change (Ammerman et al. 1998:15). Resources may include the congregations' members (including their level of skills and theological and secular education), "its money, its buildings, its reputational and spiritual energies, its connections to the community, and even its history" (Ammerman et al. 1998:15). The way in which people are able to draw on such resources as are available to them through their congregation impacts the process of change, and the extent to which change may occur. Finally, the process frame includes the way in which people in congregations interact with each other. This may include how they worship, study the Bible, and make decisions. Clergy play an important role in this process. These congregational factors intersect with participants' self-perceptions of how or why they have experienced change. These frames also may be applied to the Zero28/ikon "community," which I have included with the congregational analysis.

With these factors in mind, I sought evangelical congregations with traditional and mediating emphases in urban and rural settings. Evangelicals are well-represented within both mainline denominations and smaller denominations, so I sought congregations from a variety of denominations. Given Presbyterianism's historical prominence, I sought traditional and mediating congregations from that denomination. Within those congregations, I sought individuals of varying ages, education levels, occupations, and gender. This ensured that my sample was both broad and deep.

Practical Considerations

Locating the congregations depended on my prior contacts within Northern Irish evangelicalism.[2] I attended a variety of services and meetings with each congregation. I asked the pastors to direct me to five to seven members of their congregation who would be willing to participate in the research. I asked them to select participants with a wide range of characteristics (age, gender, education, social class, and representing a variety of theological and political beliefs). Participant observation and semistructured interviews with people in the congregations generated the raw data on which the analysis is based. Names have been changed to protect confidentiality. Ages and occupations are included in order to give some indication of the potential resources (such as skills and education) individuals may draw on.

Analysis of the Congregations

Rural Presbyterian

My traditional evangelical contacts directed me to the pastor of two Presbyterian congregations in nearby villages in rural Co. Antrim. I talked with people from both congregations. The larger congregation consists of 115 families and is situated in a town that is about 90 percent Protestant; the smaller congregation consists of 80 families and is situated in a mixed town. The religious "ecology" surrounding the two congregations is quite different. In the case of the larger congregation, it is the only church in the village. The smaller congregation is one of several Protestant churches and a Catholic chapel in the village. Five participants were from the larger congregation; while two were from the smaller congregation. Traditional industries were agriculture and fishing, but now a number of residents travel to larger towns for work.

The culture or ethos of both congregations is evangelical; for the larger congregation this is especially so. A typical Sunday morning service combined the traditional Presbyterian use of the psalms with the use of the modern mission praise hymn book and the New International Version of the Bible.[3] It also included a children's lesson. The pastor is an energetic leader who is well known in the villages. He is also a prominent Orangeman who speaks publicly on political issues, including his opposition to the Belfast Agreement. When he spoke of the Belfast Agreement, he objected to it on both religious and practical terms. He participated in the long march from Londonderry/Derry to Portadown to publicize the rights of victims and the families of victims.

I conducted five interviews with members of the larger congregation in March 2003. I was surprised that only one had a traditional identity. Three had pietist identities. They said they were not all that interested in politics and that their social activities were largely confined to the congregation. The fifth participant was a mediating evangelical. Given that I had selected the congregation in hopes of finding people with traditional identities, I contacted the pastor again and asked if he could direct me to two additional people. The pastor was surprised when I told him that three of the five people I had talked to had voted in favor of the Belfast Agreement.[4] He put me in touch with two people from the smaller congregation, with whom I talked in May 2004. They had traditional identities.

The table 4.1 summarizes the participants' demographic characteristics, identities, and whether or not they perceived change in themselves over the last few years.

Table 4.1   Rural Presbyterian Church

Name	Occupation	Age	Evangelical Identity	Perceived Change?
Sarah	Office worker	27	Pietist	Yes
George	Small businessman	74	Pietist	No
Millie	Retired teacher	74	Mediating	No
Greg	Student	16	Pietist	Yes
Jill	Manager	32	Traditional	No
Adam	Manual laborer	33	Traditional	Yes
Andrea	White-collar professional	28	Traditional	Yes

Change

Four of the participants perceived changes in themselves.[5] Greg and Sarah moved in the direction of pietism or privatization. They said they read their Bibles more, pray, and are increasingly involved in church activities. They both linked their spiritual growth to the guidance and enthusiasm of the pastor and the encouragement of others in the congregation. They conveyed the sense that as these strictly religious aspects of their identity became more important, political concerns faded ever more into the background. They did not draw on aspects of traditional evangelicalism in their constructions of identity. Greg explicitly rejected traditional evangelicalism in his move to pietism. Sarah did not articulate beliefs associated with traditional evangelicalism (such as the proper relationship between church and state); she said that such things were of no concern to her.

Adam and Andrea, on the other hand, said they have changed their political behavior. Both are from the smaller congregation in the mixed village. The pastor directed me to Adam and Andrea when I contacted him again and asked if I could talk to people who would be likely to hold (traditional) views similar to his own. They both now support the Democratic Unionist Party (DUP) rather than the Ulster Unionist Party (UUP). This change was driven by political events, and by what they perceived as the UUP's poor performance. While Adam and Andrea did not always frame these political changes in overtly religious terms, they perceived the changes as having a strong religious or moral dimension. There were liars and terrorists in government, and it contradicted their religious beliefs to support the party that had made that possible. As such, political events had the effect of crystallizing and reaffirming the importance of their traditional evangelical identities. In turn, their religiously informed reaction had direct consequences for their political behavior in that they decided to vote for the DUP.

Adam had a conversion experience in the late 1990s after a serious injury. He said that his conversion brought with it a whole new spiritual way of looking at the world. He felt that God's presence, and the support of others in the congregation, helped him to cope with the traumas associated with his injury. Beyond this dramatic spiritual change, Adam said, his beliefs about Northern Irish politics have changed. He has consistently opposed the Belfast Agreement and a united Ireland, but he previously would have voted for the UUP. Not now:

> Over the years I've been with the Ulster Unionist Party and Tony Blair and Bertie Ahern and Bill Clinton and all the rest. The lies and

deceitfulness that's been told and done—I wouldn't trust any of them. So from that point of view I'd be very skeptical now....I would always vote for the DUP now. Because I believe...that they've come out...[and what they've said] has turned out to be right. And they are fighting the unionist corner. Some of the things they would say and do, I wouldn't agree with to be honest. Sometimes I think Ian [Paisley], he speaks before he puts his brain in gear, but I would give him a fool's pardon in that. He'll never change. [May 18, 2004]

Such narratives contribute to the understanding of how identity change takes place. They allow us to link individuals' self-perceptions of change with factors that are specific to their congregations. Adam and Andrea experienced dissonances between the social order and their individual habitus, which led them to reaffirm their traditional evangelical identity. Adam and Andrea linked specific political events with moments of intentionality in which they decided that they could no longer vote for the UUP. For them, this was a religiously informed decision made in protest over the immoral way in which the Belfast Agreement had been negotiated and implemented. Greg and Sarah, on the other hand, adopted pietist identities, becoming more involved in apolitical activities with their congregation. The "ecology" or geographical locations of the congregations may have had some impact on the directions of change in the participants. For instance, the isolated nature of the largely Protestant village in which Greg and Sarah went to church may have contributed to their pietism. The "culture" of the congregation in the larger town is very church-centered. Much of village life revolves around the congregation, from youth groups to community events in the church parking lot. In a context like this, the political world may seem very far away—even with a pastor who is not shy about making his political views known.

Giving Meaning to Activism

The social lives of the people I talked with were centered around their local congregation. Congregation-centered activism reinforced the pietist tendencies of people such as Sarah, Greg, and George. Of the three people who participated in "political" organizations outside of the congregation (Jill, Adam, and the pastor), all had traditional identities. Jill plays in a flute band (which she considers more a "cultural" than a political activity), while Adam belongs to the Orange Order, the Royal Black Preceptory, and the Apprentice Boys of Derry.

The pastor is involved with the Orange Order and the Evangelical Protestant Society.

Participants believed that they could make their small villages better places through their service in the church. Many said that secularization, particularly in urban areas and amongst the young, was a big challenge facing the churches. They gave meaning to their social activism by focusing on how they could impact the local context. What was most important was preaching the Gospel. But when they started to consider the wider context, their outlooks became grim. There was a sense that the Protestant community and the churches had been deceived and marginalized. For instance, when I asked the pastor what he thought of the Presbyterian Church in Ireland's (PCI) denominational activities, he said the mainline churches were marginalizing themselves by selling the government line:

> I feel that the...mainstream Protestant churches must accept part of the responsibility for the position that we find ourselves in now....I thought it was just crazy the way our church [acted during the debate on the Belfast Agreement referendum]....I feel our church should not have got into the yes/no division. I believe our church should have stood out on and made statements on issues as regards the Belfast Agreement....For example, I felt our church should have come out clearly that you do not reward evil doers. You do not let people [out of prison] early...and put them into ministerial positions to buy off a terrorist organization. [March 18, 2003]

There was also a sense that no matter what Protestants said, they would not be listened to. As such, it was very difficult for people to find meaning for activism beyond their focus on their congregationally centered local activities. This was even the case for Jill, who had a traditional rather than a pietist identity:

> The Labour government in particular always seems to lean towards the nationalists, and Tony Blair personally seems to be very fond of them, so he does. And he doesn't seem to think there's any problem with going in a war against terrorism [in Iraq], and then letting terrorists run the show in Northern Ireland....Because...they held their guns and it's like, if you'll do this for us, we'll give you some. And then they hold onto that, and again it's the next bit and the next bit. But they'll never give [the guns] up because they're getting too much out of it. [March 19, 2003]

These narratives allow us to link the way in which people give meaning to their social activism with wider social and political processes. For instance, people with pietist identities or who have moved in a pietist direction gave meaning to their social activism by placing it in the local context. They felt that they could impact their small villages for good. The best way they could do that was by converting others to evangelical Christianity and by encouraging other believers. Evangelical pietism would be included under "privatization" in Todd's typology (2005). Both Mitchell (2003) and Todd argue that privatization is an option that has been increasingly taken up by Protestants as they react to the changes brought about by the Belfast Agreement. The implications of privatization include public indifference to politics and low voter turnout. Evangelical pietism is reinforcing the wider Protestant trend of privatization. By withdrawing from society and politics, privatized Protestants or pietist evangelicals do not contribute to outright conflict. However, they also do not contribute to conflict transformation. Brewer (2003b) argues that this sort of withdrawal is one of the most significant impediments to the transformation of the conflict in Northern Ireland.

In this congregation, participants with traditional identities also gave meaning to their social activism by placing it in the local, congregational context. But they were more likely to be involved in political or cultural organizations outside of the congregation. The pastor, in particular, thought it was important to speak out publicly about what he thought was wrong politically. When these traditional evangelicals thought about society beyond their local area, they felt pessimistic, deceived, and marginalized. For Adam and Andrea, a direct consequence of this was reaffirming aspects of traditional identity and changing their political allegiance to the DUP. By voting for the DUP, Adam and Andrea are supporting a party that has have "adapted" to sociopolitical change. For instance, the DUP affirms aspects of Protestant identity that remain in tension with its participation in some aspects of the Belfast Agreement. Todd (2005) has argued that a majority of Protestants similarly have retained core elements of their identity and "adapted" to the new order. This has meant that they at best give "grudging acceptance of the new institutions" and apply "an inappropriate logic to the institutions of the settlement" (Todd 2005:449). Traditional evangelicals are reinforcing these wider Protestant trends and providing the wider Protestant community with religious and moral justifications for doing so. However, this does not necessarily signal a fundamental rejection of all the principles of the Belfast Agreement. The adaptation of identities (even selectively) is less likely to reinforce division and conflict than the reaffirmation of oppositional identities.

Urban Free Presbyterian Church

The urban Free Presbyterian congregation serves a Protestant working-class area of the city, and it conducts regular evangelistic outreaches in the neighborhood. There are numerous other Protestant churches nearby, many of which participate in an ecumenical council. The Free Presbyterian church does not. The congregation itself is mixed in terms of age and socioeconomic background. Some of the people I talked with traveled from outside the immediate area to attend the church.

The congregation was founded about 40 years ago with just eight people and has grown to the point where about 300 attend services on a Sunday morning. The "culture" of the congregation is typical of Free Presbyterianism. Sunday morning services feature the singing of older hymns and psalms, and the King James Version of the Bible is used as a point of principle. Members turn out in their "Sunday best," with women wearing dresses and hats. There are activities at the church almost every day of the week, including seven prayer meetings per week. Prayer meetings are deemed vital to the life of the congregation and are promoted at Sunday services. I attended a Wednesday evening prayer meeting that featured long prayers for the "unsaved" and those in ill health. The pastor also prayed that God would take away the prestige and identity that the government had bestowed upon the terrorists in Northern Ireland.

The congregation has grown especially rapidly since the Belfast Agreement, so much so that an extension had to be added to the church building. The pastor estimated that the congregation had grown about 25 percent over the last five years (Interview, September 22, 2003). The pastor is well-respected in Free Presbyterian circles and is known for his articulate presence in the media. He makes it a priority to speak out about what he perceives as social and moral ills. One member of the congregation told me that the pastor had been physically assaulted for publicly speaking out about his beliefs.

I conducted the interviews between September 2003 and April 2004. Everyone I talked to had traditional identities. This was expected, given the usual association of Free Presbyterianism with traditional evangelicalism. Three of the six perceived changes in themselves. These changes were all in a traditional, Calvinist direction

Table 4.2   Urban Free Presbyterian Church

Name	Occupation	Age	Evangelical Identity	Perceived Change
Michael	Self-employed laborer	47	Traditional	No
Sandra	Professional	39	Traditional	No
Joe	Retired manual laborer	81	Traditional	Yes
Jacob	Professional	58	Traditional	Yes
Helen	Homemaker	71	Traditional	Yes
Zack	Semiretired businessman	61	Traditional	No

and occurred prior to the time of the Belfast Agreement. Table 4.2 summarizes the participants' demographic characteristics, identities, and perceived change.

Change

Three of the participants perceived changes in themselves. Jacob talked about theological change, in that he came from a Baptist background but became convinced of the superiority of the Presbyterian form of church government while he was in his early 20s. He said that this followed on from his conversion experience. Helen and Joe, on the other hand, talked about changes in their political behavior. They both changed their allegiance to the DUP a number of years ago. Helen attributed this to the preaching of Ian Paisley. Both Helen and Joe believed that other unionist politicians were not standing up for their people. Like Adam and Andrea from the rural Presbyterian congregation, Helen and Joe framed political change in religious terms of morality and deceit, good and evil. They wanted their politicians to be guided by the Bible. Both Helen and Joe's changes occurred in an uneasy political context, as the Troubles were beginning to escalate at that time. These changes parallel how Adam and Andrea's political behavior has changed in the current, uneasy political context. As such, political events had the effect of crystallizing and reaffirming the importance of their traditional evangelical identities. In turn, their religiously informed reaction had direct consequences for their political behavior in that they decided to vote for the DUP.

Helen is a 71-year-old homemaker who has lived in the area of the city near the church all her life. After she was converted, she and her husband attended a Nazarene church for about 15 years. She started attending Ian Paisley's church, Martyrs' Memorial Free Presbyterian, around the beginning of the Troubles. She and her husband began to attend their current Free Presbyterian congregation about 25 years ago, because it is closer to their home. As political events unfolded, Helen began to make more links between her religious and political identities. She reaffirmed her traditional evangelical beliefs, whilst at the same time deciding to go to Paisley's church to hear his *political* messages. When asked to elaborate on her decision to start attending Martyrs' Memorial, and how that impacted her thinking, she said Paisley's church was the only place she believed she could go to hear the "truth" about politics:

> At that time there was so much going on.... 1972 was the year of the Le Mon bomb, the Bloody Friday, the Abercorn. There was so much happening that you wanted to hear firsthand talking about it because...you couldn't believe what you read in the papers. So I knew I wanted to hear somebody who you knew was going to tell you the truth about the situation.... It wasn't to hear Dr Paisley preach, but I wanted to hear what he was thinking and what he was doing and what was being done about the situation, which he did. I mean, he told us what the government was planning to do. He made no secret—everything he knew he told to his congregation.... When that started his church was packed, with non believers as well as believers coming to hear what Mr. Paisley would say, cause they knew he would tell them the truth. And he was reviled for it in the papers. The papers reviled him because he told the truth. [April 6, 2004]

Joe is an 81-year-old retired manual laborer who was raised on a farm. He converted in his mid-20s, after his mother's death. His father had died while he was still in his teens. Like Helen, he linked changes in his political outlook to the experience of the Troubles. Joe said Catholics' desire to dominate the country led them to violence, which paid off in political gains such as the voting franchise (and eventually, with the Belfast Agreement, putting terrorists in government). He regarded this as immoral. Joe said he started to vote for the DUP early on in the Troubles and linked this with the failure of other unionists to defend their people's interests. He also said he continued to vote for the DUP because of the Christian commitment of individual DUP politicians, such as Nigel Dodds.

These examples contribute to the understanding of how identity change takes place, providing insights on individuals' self-perceptions of change. They do not allow us to link those self-perceptions with factors specific to this Free Presbyterian congregation, because these changes occurred before these people attended this particular congregation. However, Helen and Joe provided examples of how traditional evangelical identities had been reaffirmed during a period of sociopolitical unrest. Their narratives demonstrated that the process of change is multifaceted, driven by wider political events, religious beliefs, and the perception that Protestants are threatened or that no one is looking out for them. The continuity between the way Adam and Andrea from the rural Presbyterian congregation described their changes and the way Helen and Joe described their changes is striking. Their narratives shed light on one of the political implications of a reaffirmation of traditional evangelical identity: a switch in support from the UUP to the DUP. In the current unsettled context, however, no changes in identity had occurred amongst the Free Presbyterians. They held steady in their traditional identities.

Giving Meaning to Activism

The Free Presbyterians were amongst the most socially active people I interviewed. They sensed that their religious views were being marginalized in the public sphere, and they were determined to take action to make their voices heard. They believed that no one was standing up for what was right, so it was up to them to defend God's laws. In their view, there is a better chance that "right religion" will be preserved within the United Kingdom than a united Ireland, and they considered aspects of the Belfast Agreement immoral (such as the release of prisoners without paramilitary decommissioning). However, their activism was not focused on those concerns. Rather, they were far more concerned about moral issues. Although some Free Presbyterians seemed to equate morality with a unionist—or more accurately Protestant—political order, that equation was tempered with the realization that a Protestant political order no longer existed and could not be regained. Now, traditional evangelicals believed it was up to them to keep the nation from swerving off the right moral course. Sandra, a 39-year-old professional, put it in these terms:

> This scripture I was reading talked about the Lord...talking to the children of Israel and he was giving them certain things they were not to do and one of them was that they were not to offer their children unto Moloch. And then it went on to say, the Lord had actually said that these people would be cut off. And then he went on to talk about those that didn't do those things but that turned their eyes away from them, they would be cut off. And that really spoke to me. That yes, Lord, I can't just sit back and fold my arms in these things....I believe that the church does have an obligation to stand up and say, well this is wrong...and this is the consequences of that. This is what is

happening because we're relaxing those laws and because people are so concerned about everybody's rights....I think the church is really asleep to that at the moment. [December 18, 2003]

The Free Presbyterians carried out their activism through congregational activities. These activities were clearly part of the congregation's "culture." They took to the streets to protest a number of issues, ranging from the opening of sex shops, the operation of a planned-parenthood clinic, the opening of pubs on Sundays, a gay pride parade, the performance of a gay choir, and the performance of the musical "Jesus Christ Superstar." Indeed, when it came to addressing moral issues in the public sphere, this congregation was more organized and active than some special-interest groups! The pastor's views on the moral condition of the land and the necessity of revival matched those of the participants. The pastor did not seem to have encouraged the participants to *change* to accept these views (it seemed they had held them already), but he sustained them in their efforts. He helped to organize and went on the pickets with them. None of the Free Presbyterians criticized his involvement in these activities. They also did not consider them "political" activities. They described their pastor as a nonpolitical clergyman. Zack, who had also attended Martyrs' Memorial for a time, contrasted his pastor to Paisley:

> I can't remember ever once anything from the platform of a political nature. That was one of the things that attracted me there. I mean if you went to Martyrs' it would be a different kettle of fish....[In this congregation] they're there to preach the Gospel and that's it....People can hold their views, and there's nothing wrong with that...but the pulpit of the Christian church is not the place for it. [April 6, 2004]

The people I talked to in this congregation were distinctive in their unity of purpose, and in their belief that their moral activism could contribute to a revival. They drew on the well-documented history of revivalism in Ulster and, in some cases, in the USA. If these evangelicals had a "political program," it was a program designed to cooperate with God in a revival. Most claimed to be optimistic that God will bring about a revival. In the meantime, they said it is their duty to pray and to continue in their moral activism, even in the face of persecution and ridicule.

These narratives also allow us to link the way in which people give meaning to their social activism with wider social and political processes. These traditional evangelicals retained aspects of traditional identity and "adapted" selectively to sociopolitical change by focusing on moral issues. They perceived themselves as nearly the only ones standing up for morality in society. Since they believed that God blesses nations that follow his laws, it was of utmost importance to them to make sure that Northern Irish laws reflected what they perceive as biblical mandates against abortion and homosexuality. This translated into street activism. They tended to think of change in negative terms, believing that they should hold fast to what they believed was true. The participants who told me that they had changed all described those changes as something that led them into firmer beliefs about theology or politics. This created a congregational culture in which prayer, moral activism, and standing firm in the face of change were valued. This culture provided the Free Presbyterians with justifications for making their voices heard in the public sphere. But it also limited what they would say, in that it would be quite costly for anyone in the congregation to change his or her identity or to dissent from the congregational culture. If they did, they might feel as if they had to leave the congregation.

Initially, I was surprised by the extent to which they focused on moral activism and seemed to neglect political issues such as a united Ireland or the Belfast Agreement. However, their focus on morality was tempered with the realization that a Protestant political order no longer existed and could not be regained—that nothing short of the miraculous intervention of God could create it. The decision to adapt by focusing on moral issues parallels the process that has taken place within the New Christian Right in the USA. As was the case with the traditional evangelicals in the rural Presbyterian church, the Free Presbyterians have the potential to reinforce wider Protestant trends of adapting selectively to the new institutions and providing religious and moral justifications for doing so. This may contribute to periodic crises, but it does not necessarily signal a fundamental rejection of all of the principles behind the Belfast Agreement. The Free Presbyterians' organized focus on moral issues set them apart from the traditional evangelicals in the rural Presbyterian church. It may be that this congregation is unique amongst Free Presbyterian congregations in its degree of focus and its culture of prayer and activism. However, its focus on moral issues may indicate the development of more "normal" evangelical politics. If traditional evangelicals do not deem it vital or necessary to focus on political questions such as a united Ireland, then this too may signal an acceptance of the new political order. In an indirect way, it may contribute to the transformation of conflict.

Rural Church of Ireland

I met the rector of the rural Church of Ireland (COI) parish at an Evangelical Contribution on Northern Ireland (ECONI) summer school in 2001. The rector's association with ECONI, as well as the reputation of the COI as a moderate, mainline, ecumenically friendly denomination, led me to believe that this parish would be located on the mediating end of the evangelical spectrum. When I contacted the rector again in the autumn of 2002, he agreed that his parish would participate in the research.

The parish comprises two congregations, which are located in nearby towns in rural Co. Antrim, with a total of 380 families: the larger congregation (320 families) is situated in a predominantly Protestant town while the smaller congregation (60 families) is situated in a mixed town. Presbyterianism is the dominant form of Protestantism in the area. There are five PCI churches between the two towns, as well as a Free Presbyterian church and some smaller evangelical denominations. Of the interviewees, six were from the larger congregation and one was from the smaller congregation.

The "culture" of the parish combined moderately high church Anglicanism with a carefully defined form of evangelicalism. Its congregational culture was impacted by its location within the COI, which does not have as many evangelicals within it as does, for example, the PCI. The COI is considered quite moderate due to its solid ecumenical links with the Catholic Church and other Protestant denominations, although its relationship with the Orange Order (including its handling of the march to the COI church at Drumcree) has damaged its moderate image to a degree. The congregational culture also was impacted by the religious "ecology" of its geographical location, with its predominance of (evangelical) Presbyterian churches. The rector said that many people began attending the parish because they had been rejected, in some way, by Presbyterian congregations in the area. This might occur if, for example, a Presbyterian minister refused to baptize children if the parents were not "born again." This parish seemed to serve as a catchall for Protestants in the area who could not identify with "hard-line" Protestantism. The extent to which the ethos of the parish was "evangelical" was not clear. The rector has introduced more modern versions and translations of the Book of Common Prayer, but he described the worship services as "traditional" (in the sense of the high church Anglican tradition). The rector called himself evangelical, but without the "baggage" that comes with being an evangelical in Northern Ireland.[6] One of the members of the parish I talked to,

Susan, did not describe herself even as a Christian, let alone an evangelical. Another, Richard, said, "I don't follow the Bible to the strict level. I wouldn't be an evangelical or anything near an evangelical. But I live a good life, I think I do" (Interview, March 17, 2003).

The rector is not a native of Northern Ireland, but has served the parish for 13 years. He would not have the same sort of public profile as the ministers in the rural Presbyterian and urban Free Presbyterian congregations. He does not make regular media appearances, for example. But he described himself as reasonably politically minded and said he has on selective occasions incorporated political messages in his sermons. He also has participated (along with members of his congregation) in social action such as publicly supporting Catholic parishioners when their chapel was picketed by other Protestants. He has promoted cross-community activities and arranged for ECONI to facilitate a cross-community training program in the area. I expected that his beliefs might reflect or have some impact on those in his parish.

I conducted interviews from November 2002 to June 2003. I expected the participants to have mediating identities and be open to change in a mediating direction. On the other hand, I thought that the usual association of traditionalism with rural areas might work against this. Five of the parishioners had mediating identities and two had pietist identities. Four perceived changes in themselves. Three said they had moved in a "moderate" direction, while one had moved toward pietism.

Table 4.3 summarizes the participants' demographic characteristics, identities, and perceived change.

Table 4.3 Rural Church of Ireland

Name	Occupation	Age	Evangelical Identity	Perceived Change?
Susan	Student	19	N/A*	No
Jean	Retired administrator	56	Mediating	Yes
Gordon	Skilled tradesman	53	Pietist	Yes
Richard	School administrator	46	Mediating	Yes
Anna	Shop worker	39	Pietist	No
Ray	Professional medical worker	60	Mediating	No
Robert	Professional engineer	57	Mediating	Yes

Note: * Susan said that she is not even a Christian, let alone an evangelical. Therefore, it is unfair to describe her "evangelical identity."

Change

Four of the participants perceived changes in themselves. Richard said that he is more relaxed about his religion now, although his core "theological beliefs" have not changed. He said this relaxation was simply a part of "growing up." Robert also said that his theological beliefs had changed, in that he had become less judgmental and clearer about some things, such as the concept of grace. He said this was a gradual process that involved talking with other people, and thinking deeply about mathematical concepts and human development.[7] Jean said she had become more moderate in response to preaching and as a result of observing the political process with a view to understanding the Catholic/nationalist position. Although they had not moved from traditional to mediating identities, reflecting on some of the religious themes emphasized by mediating evangelicals had strengthened their religious identity and motivated their activism. For instance, the strengthening of aspects of their mediating identity instilled in them a commitment to doing "small things" at the local or personal level that would contribute to better community relations. These activities ranged from participating in cross-community activities outside the parish, supporting the parish's (cross-) community carol singing at Christmas, and supporting Catholic parishioners during a protest by other Protestants at a nearby Catholic Church.

Jean is a 56-year-old retired administrator who has been attending services in the parish for about 18 months. She grew up in the area, was raised a Presbyterian, and has attended PCI churches for most of her life. She attended another COI parish immediately prior to attending this parish. She describes herself as "actually Presbyterian" and is still a member of a Presbyterian church. However, she said the preaching, friendliness, and hymnody in the COI had changed how she feels about "others and politics." When asked to compare the COI and the PCI she says:

> [P]ossibly, and this may be more to do with the individual churches, but... possibly the Church of Ireland... talk more about how you feel towards and act towards your neighbor. I think that has struck me a bit. They're very strong on love your neighbor and therefore that has got to affect how you feel about others and politics.

This theological emphasis on "love your neighbor" coincides with what Jean perceives as more tolerance in her politics—although that tolerance is tempered with awareness that in other ways she might be "more extreme":

> I've become more tolerant and...[with] devolution...we had an opportunity to see how the different political parties...behave...and how responsible they are, and I've become...more impressed with the SDLP [Social Democratic and Labour Party] than I would have been a few years ago. But also a bit more impatient with some of the more extreme...DUP's and Sinn Fein. [March 15, 2003]

Jean links her tolerance for the SDLP with her observations that they "care more about...the issues that affect people rather than about their own party" and that "they're not always asking for more and more concessions." As far as her difficulties with the DUP are concerned, she says that that is because all they do is criticize without proposing solutions. In the case of Sinn Fein, she does not trust its links with paramilitaries.

Gordon is a 53-year-old skilled tradesman who has been a member of the parish for 10 years. He was raised in the area and has from his youth attended a COI in a nearby large town. He left that parish when a new rector began introducing changes in the youth ministry that, as a youth leader, he did not agree with.

Gordon voted against the Belfast Agreement. He would accept a united Ireland if it was brought about democratically, but he would be "saddened" by it. Gordon is a member of the Orange Order and supports the UUP. He describes himself as politically active when he was younger and he was a member of the Ulster Defence Association (UDA) before it became a paramilitary organization. Now, he perceives himself as becoming more spiritual and less interested in politics. He said this change has been part of a religious transformation that occurred largely outside the context of his parish. Indeed, the changes he has undergone have led him to disapprove of the way his rector approaches religion and politics:

> Over the past four or five, 10 years probably, less and less politics cause I've seen the way it's went...I found, you know, myself going to Bible classes, reading tracts... we had...[a] curate in [the town]...and he played football, and I could talk to him as I played football myself...and he gave me a couple of books to read....I said to him...I attend church on Sunday...I don't feel I'm doing anything wrong. I'm happily married, I have a family and all now, I says there's something still missing in my life—I couldn't—to this day I still can't say yes, that's what's missing. So he gave me a couple of books to read from theologians and

that, easy reading, then I started thinking, he talked to me about it, and then I started going to Bible classes and study things and probably just that way, no great big, some people I'm speaking to it's flashing lights they've seen and this that and the other thing. Which is fair enough, maybe they have, I don't know, but I always felt that I wasn't that far down in the gutter. People maybe who are drunkards or drug addicts or sleeping rough and that and have amazing conversions and...mine was a gradual thing, so I feel more spiritually involved now than politically involved....My views I keep them to myself personally or as I say I don't discuss them much even with my family.

After these comments, I asked Gordon if changes in his theology affected what he believed politically. He said:

Well it would yes...now if you had asked me about 12 years ago about the Orange Order and about parading, if there had been a parade every day of the week I would have been out there, but when I sit down and think about it, to me it's all wrong because we're bringing thousands out into the streets...we went home then [and] all the yahoos who had drink in them...next thing started stoning the police, they were throwing things, and then you had two or three days riot, you know, our own police officers were getting hurt. Our own people were getting hurt....But I mean 12 years ago I seen nothing really wrong with that. [March 13, 2003]

These narratives contribute to the understanding of how identity change takes place. They allow us to link individuals' self-perceptions of change with factors that are specific to their congregations. Jean as well as Richard and Robert experienced dissonances within their individual (Christian or evangelical) habitus. Reflecting on what it meant to lead a Christian life produced a shift in the way they expressed their religious identities. As such, they focused on aspects of their religious identity that allowed them to assimilate to changed social and political circumstances. They drew on resources within mediating evangelicalism that told them that it was their religious duty to contribute to peace at the local level. Jean talked at length about how her switch to the COI had impacted this process, contrasting it to her experience in Presbyterian churches. This provides us with some idea of how this parish's "culture" (including the preaching of the rector, worship services, and the parish's ecumenical activities) creates a climate that enables individuals to negotiate change in this way. Gordon provides an example of an evangelical following a pietist trajectory; a consequence of this has been withdrawing from unionist politics. He linked this change with religious experiences outside of his parish. It was striking that there were no examples of traditional identities amongst those I talked to, given the association of traditionalism with rural life. However, this may have to do with the parish's "culture" and the local religious "ecology." The parish culture promotes an enthusiasm for pluralism that is a mark of mediating evangelicalism. The local religious ecology is such that the traditional evangelicals in the area would be well served by various Presbyterian and smaller evangelical churches.

Giving Meaning to Activism

The COI participants' social activism was not like that of the Free Presbyterians, whose congregation operated like a well-organized special-interest group. Nor were they likely to be involved with a special-interest organization, such as ECONI. Like the rural Presbyterians, some of them felt that the Protestant viewpoint had been marginalized, or that there was little they could do to impact society or politics. However, they felt that as Christians, they should still do their part to make Northern Ireland a more peaceable society. They felt encouraged by their rector and others in the congregation to do this. The main strategies that they named were participating in cross-community or ecumenical activities, or simply doing "little things." Several noted that their parish had "made the effort" to instigate or participate in cross-community or ecumenical activities in the town. In contrast to the rural Presbyterian church and the urban Free Presbyterian church, most viewed cross-community or ecumenical activities positively.[8] This reflects the mediating evangelical belief in the idea of tolerance for religious and cultural pluralism, as well as a conviction that cross-community activities will in some way contribute to the transformation of the conflict.

For example, Robert cited several cases when the parish encouraged cross-community activities or participated in "political" acts. These included cross-community carol singing, showing support for a Catholic parish in the area, and inviting ECONI to facilitate a cross-community study group. A 57-year-old engineer, he has been member of the parish for about 30 years. He described how the parish organized cross-community carol singing in their church:

> One of the things that I pushed us to do in the parish was to have a Christmas carol service for the village in our church. You know, Protestants and Catholics come to it. Now it may not sound a big thing but in a village [like this], it's a step forward. And one of the

things I noticed was that an Orangeman who's in our choir came along to the service. That was him, sort of joining in, carol singing. There's nothing very dramatic about it, but at least he came. So I would like to think the church [could contribute examples like that].

Later, he provided more details about the carol singing. This narrative demonstrates his belief that the "little things" matter:

One of the decisions we made was that we wouldn't have a carol *service*, in our church, for the community. We would have carol *singing*. Now I have no doubt in a couple of years' time it will be a carol *service*. But...it's just in case—there are some very, very hard-line people [in the village]. Not many in our church, but people who would stir things up....So we felt that if anybody came and stirred up we'd say: we're singing Christmas songs together....Where is this terrible thing, you know?...But once it's accepted and people come along it's sort of increasingly accepted....I have a very strong belief, it's one of my core beliefs, that you do what you can. And I just think that over a long period, these little things do matter. [March 14, 2003]

Robert also described how he had attended services at a Catholic Church when a group of Protestants were protesting outside it. For him, this was a way to show support for the Catholic worshippers and to take another small step forward. The parish also attempted to build relationships when it invited ECONI to facilitate a cross-community discussion group in the area. It was held about 18 months before I conducted my research in the parish. It was held in the home of a COI parishioner, rather than in the church building, so that Catholics would not feel as if they "were coming onto Church of Ireland territory." The rector said the attendance was 17 or 18 people, split fairly evenly between Protestants and Catholics. All but one of the Protestants was from the COI. They had invited Presbyterians to attend, but only one came. Robert was the only parishioner that I talked to who mentioned attending this study group. He said he found it stimulating, although he would have liked more Catholic and Presbyterian participation.

In addition, Anna and Richard said that cross-community contacts outside of parish or church-related activities were significant ways to build better relationships with Catholics. This reflected a determination to make cross-community contact a part of their everyday lives. They said that participating in cross-community sports promoted peaceable living amongst Protestants and Catholics. In Northern Ireland, rugby is one of the few sports that both Protestants and Catholics play. Richard coaches youth rugby, while Anna's two sons are involved in the sport.

In sum, people with mediating identities gave meaning to their social activism by doing the "small things" within their parish or in their everyday lives. Their activism was not always congregation-centered, but the congregation was a vehicle that equipped them for small, everyday acts that they considered socially (and perhaps even politically) significant. For those who experienced change, it was not a change from another religious identity to a mediating identity. Rather, they began to focus on aspects of their mediating identity that they felt were important, such as tolerating or encouraging pluralism or promoting peace through cross-community action. This focus was reinforced by a congregational culture that encouraged cross-community action. Even Anna and Gordon, who had pietist identities, adopted aspects of this everyday approach into their thinking. Their pietism might have made them shy away from outwardly "political" acts, but they perceived themselves as doing the "little things" on a day-to-day basis.

The process whereby mediating evangelicals are making particular aspects of their identity more important is part of what Todd calls "assimilation," in which people reshuffle aspects of their identity, rejecting some aspects of it and/or placing other aspects closer to the center of their identity. For these mediating evangelicals, the consequences of assimilation are increased involvement in cross-community activities and an everyday awareness of what they believe they should be doing to promote peace. Todd argues that a minority in the wider Protestant community have assimilated in a similar way, for reasons ranging from a desire for stability (such as recognition of the practical or economic benefits of the new sociopolitical situation) to religious reasons (such as the experience of working with Catholics in civil society organizations). These Protestants tend to be among the wealthy and highly educated. Mediating evangelicals reinforce this trend, providing a religious and moral justification for it. Their position within the Protestant community *as evangelicals* gives them a significant platform to make their case to the wider community. To the extent that cross-community activities and everyday "small steps" produce their desired effects, mediating evangelicals contribute to the transformation of the conflict.

Urban Presbyterian

I came to the urban PCI through the 2001 ECONI summer school. I met Kelly, who attends this congregation, at the summer school. Kelly is active in ECONI, and I interviewed her about the organization in 2003. As my research progressed, I remained in contact with Kelly and she offered me overnight accommodation in her home when I came to Northern Ireland for research. I accepted her invitation and attended church with her on several occasions. This convinced me that the church she attended fit within the theoretical framework I had developed. So I asked Kelly to put me in touch with her pastor, and he agreed to facilitate the research.

The congregation is large (912 families) and wealthy (it ranks near the top of the PCI in terms of congregational giving). It has been growing over the past decade, attracting more young families and young people. The head pastor attributed this in part to a concerted effort to "focus much more on youth work, children's work, and to make our whole approach to community more friendly, welcoming, and inclusive" (Interview, May 19, 2004). It is located in a wealthy area of the city, although there are working-class estates within walking distance. There are other Protestant denominations nearby, many of which participate in an ecumenical council. This congregation does not participate on the council, but it is for practical rather than theological reasons. Members of the congregation participate in a number of other "peacebuilding" and/or cross-community initiatives in the city. The congregation recently appointed a peace agent to focus on promoting peacebuilding activities amongst members of the congregation.[9] Several prominent politicians are part of the congregation. The congregation also has links with mediating evangelical organizations such as ECONI and Evangelical Alliance. Both the head pastor and the associate pastor described the congregation as evangelical.

There are two worship services at the church on Sunday mornings and a large Sunday school class. One of the Sunday morning services is quite formal, with the ministers wearing robes, and the other is less formal; the Sunday evening service is still less formal. The congregation sings a mixture of contemporary praise songs and classic hymns and uses several modern translations of the Bible. The congregation does not follow a liturgical calendar, but several months in advance the ministers set a Scriptural program from which they will preach.

The head pastor was active in ECONI around the time of its foundation and has a relatively high public profile. The associate pastor has strong links with evangelicals in the Republic of Ireland and has participated in Catholic-evangelical initiatives. They said they believed that the Gospel has political implications; as the associate pastor put it, "I think it was [Desmond] Tutu ... who said people who say religion

Table 4.4   Urban Presbyterian Church

Name	Occupation	Age	Identity	Change
Kelly	Scientist	48	Mediating	Yes
Brad	Medical professional	59	Mediating	Yes
Melissa	Community worker	41	Mediating	Yes
Ken	Retired professional	60	Mediating	Yes
Darin	Legal professional	44	Mediating	Yes
Winslow	Semiretired executive	68	Mediating	Yes

and politics don't mix, I don't know what Bible they're reading" (Interview, May 19, 2004). For these pastors, it meant walking a fine line between ministering to the congregation and drawing out the political implications of the Gospel, when appropriate.

I conducted the interviews in May and June 2004. Of the six people from the congregation I spoke with, all had mediating identities. This did not come as a surprise, given the reputation of the church. In addition, I think that the head pastor purposefully directed me to people who were active in peacebuilding activities outside of the congregation. The people I talked with were involved in a variety of community initiatives, ECONI, and the One Small Step Campaign.[10] Everyone I talked to had perceived changes in their beliefs or identities. Unlike the Free Presbyterians, they were positive about change and saw it as integral to their development as Christians. Table 4.4 summarizes the participants' demographic characteristics, identities, and perceived change.

Change

Like the mediating evangelicals in the rural COI, these evangelicals had not moved from traditional to mediating identities. Rather, they had begun to draw on resources within the mediating identity that allowed them to assimilate to changed social and political circumstances. For instance, Winslow said that he had become more accepting of others. He described this as a life-long process that included leaving what he called a fundamentalist Brethren church, prayer, and interaction with people of other religious beliefs. Ken said that he had become more moderate, and that it was a gradual process that involved growing older and observing the political developments of the last 30 years. Melissa said she had become more open. She attributed this to having lived in the Republic of Ireland and in a cross-community religious retreat center for a time, self-consciously participating in dialogue with people from different points of view. Darin said his changes came through watching the way other Christians set a good example in the public sphere, playing rugby, participating in cross-community dialogue, and taking on the lessons learned through organizations such as ECONI and One Small Step. Kelly said her changes were related to living abroad and to participating in a number of evangelical and cross-community groups (ECONI, the Irish School of Ecumenics' Moving Beyond Sectarianism Project, the Irish Christian Studies Centre, and youth work). Like the mediating Anglicans, these evangelicals perceived these changes as a strengthening of their religious identity. The strengthening of their religious identity in turn motivated their activism.

Nearly all of the mediating evangelicals I talked to in all of the congregations said they usually supported the UUP or the Alliance Party (even if they did not support the Belfast Agreement or approve of the way it had been implemented). Brad, however, provides an example of a mediating evangelical who had—for the first time in his life—supported the DUP in the last election. Like Adam and Andrea in the rural Presbyterian congregation, he linked this change to dissatisfaction with the way the agreement had been implemented. However, he did not frame his dissatisfaction in terms of good and evil, morality and deceit, as Adam and Andrea did. Although his political behavior had changed in response to events, that change was not reinforced by his religious identity. Rather, he said he had become more inclusive in his religious beliefs and was convinced that the churches should be emphasizing themes such as Christian peacebuilding and Protestant/Catholic dialogue. He said that he now considered himself a Christian rather than a Presbyterian.

The participants shared with the rural Anglicans a conviction that it was their religious duty to "do something" to contribute to peace, and their urban location allowed them to tap into the resources of an array of evangelical organizations that addressed the issues they thought were important. They have become involved in organizations dedicated to peacebuilding, and they have tried to encourage their congregation to do more in this area. Even so, the participants often felt like a minority within their congregation, saying that most of their coreligionists did not see peacebuilding as a priority in practicing their faith.

Kelly grew up on a farm in Co. Tyrone. She attended a Methodist church and was involved in a number of interdenominational evangelical organizations as a child. She went to university in England, where she joined a Baptist church. When she returned to Northern

Ireland, she attended a Baptist church in Belfast for about 12 years. Then she attended a Presbyterian church for 10 years after becoming involved with a house group based in that church. She then moved from that church to this congregation, where she has been about five years. Kelly said she first began experiencing changes in the way she looked at Northern Ireland (theologically and politically) when she went to university. When she returned, she became involved in a number of organizations and groups that helped her to change:

> I was involved in the Irish Christian Studies Centre here. This organization did set itself up to help us think Christianly about every and any issue. About how we could put our faith into practice in dealing with life and society.... We were actually just collections of people discussing issues and maybe doing practical work on unemployment, which in the 80s was a major problem in Northern Ireland, and on Third World affairs, and on the situation here in Ireland.... Now the only one at the moment that's continuing...is a study group looking at what's happening in Ireland and Northern Ireland.... Now we are kind of a fellowship group.... We've puttered along doing our little things, our little bit, and I think for the people involved it was a journey that changed us and that maybe had a little bit of an impact along the way.... [And] ECONI, what can I say about ECONI? Well, ECONI is people with a vision and a group of people who are able to take that vision and bring it to pass in ways that are relevant to the society.... I think just being part of that has helped me to understand how organisms can work in powerful ways where they are sensitive and responsive to their environment. To the people they're meeting out there on the street in the churches, but also across a...broader range than evangelical churches and even further in politics and society at large. I think in terms of living out the good news in the biggest sense of the word they...have done that in groundbreaking ways, actually. [June 10, 2004]

In sum, these Presbyterians experienced dissonances within their individual (evangelical) habitus. Like the rural Anglicans, they focused on aspects of their religious identity that allowed them to assimilate to changed social and political circumstances. They drew on resources within mediating evangelicalism that told them to take cross-community activity and peacebuilding seriously. Their urban location, with its networks of cross-community and evangelical peacebuilding organizations, provided them with more resources to do this than were available for the rural Anglicans. Their relative wealth and high levels of education provided them with additional resources. However, Brad demonstrated that religious change is a complex and multidimensional process and does not necessarily translate simply into party political behavior. For instance, he voted for the DUP for solely political reasons. The congregational culture was one in which people had the space to experience change as a positive thing and, on occasion, to reflect on the political implications of the Gospel.

Giving Meaning to Activism

Participants were active in the congregation and in special-interest organizations and tried to live out their beliefs in small, everyday steps. In this regard they were similar to members of the rural COI, although the COI parishioners did not have a high rate of participation in special-interest organizations. This may reflect the greater availability of special-interest organizations in an urban as opposed to a rural setting. It may also be that this sample was biased toward activists, because the head pastor knew I was researching religion and politics and thought it would be best that I talked to people who were actively involved. However, the participants believed that other evangelicals were not focusing on the aspects of religious identity that they thought were important. They even criticized their own congregation in this regard, despite its recent efforts to appoint a peace agent and to set up a peace group. They believed that the people within the churches either were apathetic or were worried about the wrong issues. This gave their activism urgency and meaning. They were sustained in their activism by a sense that someone had to do it, and by encouraging signs that they were helping others to change. This congregation stood out amongst the others in that everyone insisted that the church was not doing enough—particularly in the areas of peace work and assistance for the poor. For them, this lack of focus is hindering the development of a peaceful society.

Melissa, a 41-year-old community worker, grew up in the local area and attended the church from her youth to the time she was married. She lived in the Republic of Ireland for a short time, as well as in other parts of Northern Ireland, attending both Presbyterian and "community" churches. She also lived in a cross-community reconciliation center. She said her congregation is not doing enough. When asked why she thought that people did not see peacebuilding as a priority part of their Christian faith, she said:

> I suppose maybe people are involved in different things and maybe some people would see mission or their church involvement...as more priority....As well as that peacebuilding can be a bit scary, you know, because you may be asked to think about your own stuff, your own

attitudes, so it can be quite challenging as well. I wonder sometimes can it be frightening, or not frightening but a bit challenging for people and that might be a reason. I suppose another reason maybe is that people have been used to living and working in middle class... Belfast, whereas I've had the opportunity to go out of that whole environment and experience very different environments. So if you had been in that one environment all your life and hadn't really mixed with people of the other community a lot, it would maybe seem a bit scary. And it's more comfortable to stay where you are, so. [May 20, 2004]

The associate pastor also commented on the difficulty of "bringing the church with us" in implementing new programs, particularly those that have to do with peacebuilding. When I asked him if he thought it would be a big challenge bringing the church along, he said the biggest hurdle was apathy rather than enmity. This echoed the observations of the congregants that the church just was not doing enough:

[It won't be] as much [of a challenge] as it would be in other places. Not as much as it might have been maybe four, five, six years ago. It's just a case of us being sure that we're going in the best direction at the right speed. I think the problem won't be hostility but will be apathy.... [For instance], there's an event this evening which we were given... about 20 places for, to go and listen to the stories of victims of the Troubles.... We probably only filled half our places.... One of the organizers was a bit discouraged with the general turnout and I was saying [to him]... that it's important for the ones who are there. You know, the one small step idea. But I think it's not that people are opposed to it but it's just not high enough on their agenda. It's apathy. [May 19, 2004]

Like the mediating Anglicans from the rural congregation, these evangelicals experienced change as positive and focused on aspects of their mediating identity that they felt were important. However, the urban Presbyterians said their congregation and the wider church were not doing enough. They felt it was up to them to press on despite this apathy, and to encourage others to join with them. In this regard, they resembled the Free Presbyterians, who complained that the other churches were not doing enough to stand up for what was right. However, the issues that the urban Presbyterians were concerned about were vastly different from the issues that the Free Presbyterians were concerned about. The urban Presbyterians concentrated on poverty and peace, whilst the Free Presbyterians concentrated on abortion and homosexuality. The urban Presbyterians had positive impressions of cross-community activities and ecumenism; the Free Presbyterians did not. The urban Presbyterians exercised their activism through networks of organizations, complaining that their congregation was not doing enough. The Free Presbyterians, on the other hand, acted through their congregation, arguing that the Free Presbyterian Church was the only denomination that was doing something. It was not clear whether the urban Presbyterian congregational "culture" reinforced their activism in the way that the Free Presbyterian congregational culture did. While the congregational culture did not *constrain* the Presbyterians, they would have liked it to provide them with more support for their activism. The mediating evangelicals I talked to conveyed a sense that their congregational culture did not actively seek to dissuade people from apathy or pietism, and that this inhibited the transformation of conflict in the wider society.

The process whereby mediating evangelicals are making these aspects of their identity more important is part of what Todd calls "assimilation." Mediating evangelicals reinforce this trend, which Todd claims is occurring amongst a minority of Protestants. Mediating evangelicals' complaints that not enough people view peacebuilding and cross-community activity as important indicate that change in this direction may be slow.

Ikon and Zero28

I did not originally plan to include ikon and Zero28 in the congregational analysis—I thought I was studying evangelical organizations when I first made contact with them. I identified Zero28 on an Internet search on ECONI; as the principal founder of Zero28, Gareth Higgins, is a member of ECONI. Judging from Zero28's Web site and informal conversations with Higgins and others, I decided to include Zero28 in my study of organizations within Northern Ireland's mediating evangelical network. However, it soon became clear that although Zero28 shared many of the concerns of mediating evangelicals and frequently cooperated with mediating evangelical organizations, it had significant differences from mediating evangelicalism. This impression was confirmed when people within Zero28 began to tell me about their involvement in ikon. Ikon, which organizes a monthly meeting at a bar in Belfast, amongst other things, seemed not quite an organization and not quite a congregation.[11] Activists in ikon and Zero28 resisted the "organization," "congregation," and "church" labels, preferring to call themselves a "community." Because ikon and Zero28 carried out the primary functions of congregations (socialization, including identity change) and organizations (nongovernmental politics), I chose to include them both in this chapter on congregations and in the following chapter on organizations. In this chapter, I concentrate on their narratives about how their identities have changed to postevangelicalism.

Ikon and Zero28 operate in Belfast. One participant said he thought that a community such as ikon *required* an urban, university town in order to flourish. As such, the "ecology" surrounding ikon and Zero28 is as religiously and culturally diverse as it gets in Northern Ireland. Most of the participants were young and well educated. I conducted ten interviews from March to May 2004. All of the participants had third-level degrees, including two with doctorates and one who was finishing a doctorate. Six were from Northern Ireland, and one each from the Republic of Ireland, Scotland, England, and the USA. As such, the demographic characteristics of ikon and Zero28 are elite and atypical. They have significant resources of skills, education, and mobility to draw upon.

Seven of the ten participants had postevangelical identities, while three had mediating evangelical identities (with significant sympathy toward postevangelicalism). I was initially surprised to find postevangelicalism within Northern Ireland, not having encountered it in my own daily living within evangelical subcultures in the USA and the Republic of Ireland. All of the participants were involved with other evangelical, ecumenical, or secular community organizations, and for half of them, their full-time work was for a church, community, or sociopolitical organization. All of the participants perceived changes in their beliefs or identities. Like the urban Presbyterians, they experienced these changes as a positive thing and perceived themselves as moving from narrow to broader standpoints.

Like the congregations, ikon and Zero28 have created a distinct culture. The public event that is most like a congregational worship service is ikon's monthly meeting. At the time I conducted interviews, this meeting was in the Menagerie Bar, a venue near Queen's University which has since closed. Now, ikon meets in the Black Box, a bar and performance arts center in central Belfast. This is intended as an event that includes liturgical and ritual elements but where people can explore religious ideas that are not normally considered or accepted within the churches. The very act of meeting in a bar is a rejection of some strands of evangelicalism that advocate abstinence or regard bars as "sinful" places. Attempts to foster community are carried out more informally, but intentionally. They meet together in their homes and in coffee shops. Their high levels of education mean

Table 4.5   Zero28 and ikon

Name	Occupation	Age	Identity	Change
Jackson	Artist	37	Postevangelical	Yes
Meredith	Community worker	23	Postevangelical	Yes
Bill	Media professional	31	Postevangelical	Yes
Carl	Doctoral candidate	30	Postevangelical	Yes
Gwen	Nonprofit sector employee	31	Mediating	Yes
Janene	Nonprofit sector employee	30	Postevangelical	Yes
Donny	Nonprofit sector employee	27	Mediating	Yes
Larry	Lecturer and author	29	Postevangelical	Yes
Damon	Medical professional	52	Mediating	Yes
Nick	Nonprofit sector employee	36	Postevangelical*	Yes

Note:* Although I have classified Nick as a postevangelical, he was somewhat uncomfortable with the term. He preferred "pre-evangelical."

that their meetings and discussions are often intellectual and philosophical, making their culture fully accessible only to elites.

Table 4.5 summarizes the participants' demographic characteristics, identities, and perceived change.

Change

Participants had moved to postevangelicalism from a variety of points on the evangelical spectrum. Six (Jackson, Meredith, Bill, Carl, Larry, and Nick) said that they had attended traditional or charismatic[12] evangelical churches in their youth, and their narratives recounted movement from traditional or charismatic evangelicalism, to mediating evangelicalism, and then to postevangelicalism. Of these, all but Nick (who is from the USA) were from Northern Ireland. The three who had retained a mediating identity (Gwen, Donny, and Damon) chose to draw upon resources within the mediating identity to explain the changes that had occurred in themselves and their new focus on issues such as peacebuilding and Protestant/Catholic dialogue. Of these, Gwen is from the Republic of Ireland, Donny is from Scotland, and Damon is from Northern Ireland. Janene, who is English, converted to Christianity as a young adult living in the Republic of Ireland and almost immediately identified herself as a postevangelical.

The participants provided a variety of explanations for why they think that they have changed, but the overriding theme of their narratives was disillusionment with evangelicalism—particularly in its traditional Northern Irish form. They felt most acutely the "dissonances" between the social order and the individual habitus and within the individual habitus, as described by Todd. The main sources of their disillusionment were the way they perceived evangelicalism supporting unionist politics and the evangelical churches' neglect of social justice and peacebuilding. For instance, of the Northern Irish participants, Jackson said he had changed because of disillusionment with evangelicalism, the impact of living for a time in an ecumenical religious community, and reflection on the Catholic Worker Movement and the nonviolence of Gandhi. Meredith said she had become disillusioned with evangelicalism and perceived a broadening of her perspective when she lived for a year in the USA. Bill perceived his changes as linked to going to university, starting to think for himself, and disillusionment with evangelicalism. Carl talked of his conversion experience as an opening up of his mind that eventually led him to construct a critique of evangelicalism as a movement influenced too much by Enlightenment assumptions. The study of philosophy and postmodernism continue to contribute to changes in his outlook. Larry said he had become disillusioned with evangelicalism, which led to changes in his identity and a commitment to social justice. He continues to be influenced by American organizations such as Sojourners and Tony Campolo's Evangelical Association for the Promotion of Education (EAPE). Likewise, Damon spoke of disillusionment with evangelicalism and the influence of Sojourners, Mennonites, and other Christian pacifists. Significantly, of the six participants from Northern Ireland, the five that came from what they perceived as restrictive traditional or charismatic backgrounds adopted a postevangelical identity. Damon, who retained a mediating identity and attends a charismatic church, still spoke of disillusionment and had sympathies with postevangelicalism. The disillusionment or dissonance experienced by the other five led them not to assimilate their identities, as mediating evangelicals have done, but to convert to a new identity altogether.

This narrative from Meredith, a 23-year-old community worker from East Belfast, captures that disillusionment. She discusses the reasons for her disillusionment, and how she perceived it as contributing to a change in her identity:

> Zero28 are these disillusioned evangelicals who don't really know where they're looking for God anymore or how to experience that. Many of them have come from a charismatic background where you're supposed to fall over in a church and that means you've experienced God. And I think it's just a way of reminding people that God is everywhere—so much of what Zero28 is about is trying to find God in our

world and our communities rather than the evangelical mindset of imposing God on our communities. [March 10, 2004]

The four participants who are not from Northern Ireland provide an interesting contrast. The American Nick moved to postevangelicalism from a point similar to that of the Northern Irish postevangelicals who were disillusioned with evangelicalism. Nick said his negative impressions of both American and Northern Irish evangelicalism changed him, along with his study of Christian mystics and Catholic thinkers, and his involvement in the anarchist movement. Janene, from England, converted almost immediately to postevangelicalism after a brief glimpse of traditional evangelicalism (she did not like what she saw). Gwen (from the Republic of Ireland) and Donny (from Scotland) retained a mediating identity but, like other mediating evangelicals, emphasized peace-building and dialogic aspects of that identity. They also felt that the institutional churches were not doing enough in these areas.

Janene, a 30-year-old nonprofit sector employee, said that she experienced a "culture shock" when she encountered the evangelical subculture after her conversion. She could not fit into that subculture for long. Janene said that she has thought long and hard about what makes people become a part of the postevangelical community. She called this an elusive search for the "X factor" that contributes to change within people. For herself, she linked it to her youth and "postmodern, postevangelical angst." But her interaction within the ikon and Zero28 community has caused her to observe people from older generations asking the same angst-ridden questions. Those observations have led her to conclude that the change is linked with disillusionment and pain:

> The X factor has something to do with the fact that at some point [evangelicalism]...no longer makes sense in terms of people's experience in reality. For me the question then is what is it that causes the fundamental trigger? Is there some common thing that spurs that change?...To me it seems to be frequently a point of hurt, pain, or anger. The spurring is a negative one and not a positive one and that is where that disconnect happens....When their life doesn't fit into that box and when questions are posed, either through relationships or lifestyle choices....Uncertainty when their general experience of the world doesn't fit...into that evangelical box which so much of the time is just saying, where are you with Jesus? Where are you with Jesus? And the same sermon is preached every single Sunday....When they can't get their minds to fit inside that box something happens. [March 11, 2004]

These narratives contribute to the understanding of how identity change takes place. Participants from ikon and Zero28 experienced dissonances within their individual (evangelical and/or charismatic) habitus. Seven of them converted to a new religious identity altogether: postevangelicalism. Disillusioned, they rejected the way old evangelical identities were associated with political unionism and the old social order. Interaction with others was important in that process. The participants who had mediating identities provided additional examples of how evangelicals have assimilated to changed social and political circumstances. The "ecology" surrounding the community may have had some impact on the direction of change, given that ikon and Zero28 are situated in Northern Ireland's most diverse university town. The participants, with their high levels of education and skills, also had manifold "resources" to draw on as they experienced change. Finally, the "culture" of their community encouraged change in radical new directions. The community at times seemed to operate like a support group for people who had been bruised by evangelicalism, providing them with the encouragement to convert to a new identity.

Giving Meaning to Activism

Participants saw social activism as a fundamental part of their lives and vital for expressing their identities. Indeed, Zero28's original purpose was to promote activism centered around peacebuilding in Northern Ireland. Like the urban Presbyterians, they believed that the churches and/or evangelicals were not doing enough, and that they addressed all the wrong issues (abortion, homosexuality, political unionism). Some of them judged evangelicalism and the institutional expressions of Christianity as so inadequate that they no longer attend church.[13] However, they were still motivated by their faith, which compelled them to be concerned with issues such as peacebuilding, the environment, and assistance to the poor. They participated in activism through events they organized themselves or through other secular and religious organizations.

Bill, a 31-year-old media professional, grew up in a large town near Belfast. He said he had "quite a few" conversion experiences over the years but now considers himself a postevangelical. He gave meaning to his activism in terms of dissatisfaction with the kind of issues the churches choose to be concerned with:

> I think for me and a lot of people...the evangelical church [takes up]...issues that I'm not convinced are as important....They get very hung up on issues like sex before marriage, homosexuality,

drinking—yet they have no problem with the church having shares in arms companies and companies that are exploiting people in the Third World. And yet statistics like...20,000 children die every year [from HIV]. It seems to have no impact on the church at all. And yet they worry about the possibility of a gay marriage....And I'm not saying we shouldn't have any morals, I'm saying that there are bigger questions that the church completely misses....I mean when you look back even in the nineteenth century when slavery was considered normal and the church supported it, most people in the church would have said, yeah, slavery is absolutely fine. Now we look back and think how could they have supported that? You kind of wonder what people in 100 years will look back and say about our generation, some of the things that we do. And I think it's going to be things like trade, its going to be things like debt. You know...the churches have been so slow to get involved in debt issues and trade justice, and even fair trade products. Cause the church in Northern Ireland is probably one of the biggest consumers of tea and coffee but they don't drink [fair trade]....So many churches aren't even interested in getting fair trade tea or fair trade coffee, they just don't see it as an issue, you know. They're more worried about young people on drink or having sex....They kind of make a hierarchy of the wrong things and they're missing the big issues. [March 17, 2004]

The postevangelicals gave meaning to their social activism by focusing on the issues that they thought the evangelical churches had neglected. In this sense, they were like mediating evangelicals. Unlike mediating evangelicals, they experienced a much deeper disillusionment with evangelicalism. Their disillusionment was so profound that it led some of them to abandon evangelicalism and the institutional churches. In Todd's terms, postevangelicals have "converted."

At first glance, the sociopolitical implications of an identity converted to postevangelicalism and an assimilating, mediating identity are the same: increased commitment to peacebuilding and social justice–centered activism. However, there are two points of divergence. First, to the extent that postevangelicals are distancing themselves from evangelicalism, they may not be able to remain engaged with it. This could keep them from contributing to change in other evangelicals. Second, the extent that they are distancing themselves from evangelicalism may enable them to push identity-formation in more radical directions than mediating evangelicals. People with identities that transcend Catholic, Protestant, or evangelical identities would seem to contribute to the transformation of conflict in Northern Ireland. But it is not clear whether this is a viable option for more than a few elites.

Conclusions

Evangelicalism's privileged relationship with power has broken down, and its position within civil society is ambiguous. These wider changes have interacted with changes in evangelical identity, four types of which have emerged: traditional, mediating, pietist, and postevangelical. This chapter analyzed how evangelical identity change is taking place within congregations and at the intersections between individuals' self-perceptions of change and their location in a particular congregation. For example, individuals drew on a number of factors to explain personal changes, including experiences abroad, cross-community or ecumenical interaction, education, observing political events, and preaching, amongst other things. These self-perceptions were set in congregational contexts, which allowed for some factors that contribute to change to be the same for all the people in the congregation. These included the ecology surrounding the congregation, the congregation's culture, the resources available to the congregation, and the process by which people in the congregation interact (including the role of the pastor). For example, the culture of the Free Presbyterian congregation reinforced individuals' commitments to prayer, moral activism, and standing firm in the face of change, whilst the culture of the COI parish encouraged individuals to work for change by participating in cross-community activities or by taking "small steps" in their everyday lives. The ecology surrounding the urban Presbyterian congregation and ikon and Zero28 provided them with abundant resources for participating in cross-community activism or peace groups—resources that were not available to individuals in the rural COI or the rural Presbyterian church. These factors demonstrated that identity change is multifaceted and complex and cannot be reduced to one factor, such as theological reflection.

This chapter also explored the implications of identity change for the transformation of the Northern Ireland conflict. Drawing on Todd's framework, it argued that traditional evangelicals have "adapted," mediating evangelicals have "assimilated," pietist evangelicals have "privatized," and post-evangelicals have "converted." Evangelicals reinforce these wider trends amongst the Protestant community. With their much-remarked-upon commitment to activism and their historic significance, they have the social and cultural resources to drive change in new directions.

Protestants who have adapted have retained core aspects of their identity whilst selectively accepting aspects of the new order. In the case of traditional evangelicals, they focus on aspects of the Belfast Agreement that are "immoral" (such as prisoner releases and the failure of the Irish Republican Army to decommission), or on "moral" issues such as homosexuality or abortion. They are likely to support the DUP—a party that also has "adapted." This has meant that they have contributed to "crises" in the implementation of the Belfast Agreement—but it does not necessarily mean that they oppose all of its provisions and are working to restore a "Protestant parliament for a Protestant people." Assimilators have critiqued traditional evangelicalism and placed other aspects of their evangelical identities closer to the center of their identities. In the case of mediating evangelicals, this has translated into a rejection of covenantal Calvinist aspects of traditional evangelical identity, and into placing Anabaptist concepts such as a high regard for pluralism and peacebuilding closer to the center of their identities. This has led them to participate in peacebuilding organizations with other Christians and secular actors within civil society—activities that contribute to the transformation of conflict. Those who have privatized have withdrawn from society and politics, the consequences of which may be general apathy or staying away from the polling booths. Privatization also implies that they would not be inclined to participate in social or political organizations that are dedicated to peacebuilding. Those who have converted have judged their old identities as harmful or irrelevant and have changed their identity altogether. In the case of postevangelicals, this has meant rejecting traditional evangelicalism and judging even mediating evangelicalism as an inadequate way to exercise their faith. The consequences of this have included enthusiasm for pluralism and participation in cross-community and peacebuilding initiatives— actions that have transformative potential. However, the way in which postevangelicals have distanced themselves from evangelicalism could keep them from contributing to transformative change in other evangelicals. Their concentration amongst the young, urban, educated elite may also limit their influence.

Chapter 5

Evangelicals and the Reframing of Political Projects

In Northern Ireland, evangelicals have been active in forming special interest organizations devoted to specific goals. However, recent social and political changes have created dissonances between the goals and discourses that were once important to evangelicals and the acceptance of those goals and discourses in the public sphere. Evangelical organizations have responded by changing their goals and reframing their sociopolitical projects. Focusing on the way evangelicals have reframed their sociopolitical projects is more useful than evaluating the extent to which they have achieved particular short-term goals. This is because the new discourses used to reframe sociopolitical projects may have the longer-term effect of changing discourses and perceptions in the wider public sphere. This chapter draws on the typology for reframing sociopolitical projects, outlined in chapter 2, to evaluate how evangelicals are engaging in the public sphere.

The chapter proceeds by outlining theoretical and practical considerations in the selection of the organizations. Then, it explores how organizations in traditional and mediating networks are practicing nongovernmental politics. This includes an overview of what the organizations do, and how they frame their activism. This is vital for understanding how changes in evangelical goals and public discourses impact wider social and political processes, including the transformation of conflict.

Selection of the Organizations

Theoretical Considerations

The selection of the organizations was theory-driven. I sought to locate organizations with both traditional and mediating identities.

Organizations' identities could be determined by evaluating their mission statements, the literature they produced, and their Web sites. I used the same criteria to classify an organization's traditional or mediating identity as I used to classify the identities of individuals within the congregations. Although there are many active evangelical organizations in Northern Ireland, there has been surprisingly little research about them. Evangelical Contribution on Northern Ireland (ECONI) receives mention in Appleby (2000), Liechty and Clegg (2001), Mitchel (2003), Brewer (2003b), and Brewer, Bishop, and Higgins (2001); the Caleb Foundation is criticized by Liechty and Clegg. These works only begin to explore how evangelical organizations are impacting civil society.[1]

Practical Considerations

Locating the organizations depended on my prior contacts within Northern Irish evangelicalism.[2] I located the three traditional organizations—the Evangelical Protestant Society (EPS), Caleb Foundation, and Independent Orange Order (IOO)—through an activist who had been recommended to me by a Northern Irish pastor working in the USA. This activist is involved in all three organizations, and he put me in touch with others in leadership positions within the organizations. I conducted a total of eight interviews with activists in the traditional network.

I located the mediating and postevangelical organizations—ECONI, Evangelical Alliance (EA), ikon, and Zero28—through my evangelical contacts in Dublin. This led me initially to the ECONI summer school, where I met staff members of ECONI and EA. I conducted a total of 22 interviews with activists in the mediating/postevangelical network. I talked to more of them because there was less overlap in the leadership positions in their organizations. ECONI and EA were also larger organizations with paid staff. From there, I read the literature produced by the organizations and attended a number of their events. These included celebratory centenary lectures sponsored by the IOO, a Thanksgiving service for the EPS, ECONI conferences and lectures, Zero28 events such as film nights and poetry readings, and ikon gatherings in the Menagerie Bar. I located people to participate in the interviews through the snowball technique, relying on the recommendations of key people within each organization. These observations and interviews provided the raw data on which the analysis is based. It includes an overview of what the organizations do, followed by narrative analysis of how the participants have reframed their social and political projects. Participants' identities have been concealed to protect confidentiality.

Traditional Evangelical Organizations and their Activities

The Independent Orange Order (IOO)

The IOO (its official name is the Independent Loyal Orange Institution) was established in 1903. The immediate cause of its formation was Tom Sloan's expulsion from the Orange Order. Sloan had interrupted sitting unionist MP and county grand master Col. Saunderson during his platform speech at the 1902 Twelfth of July celebration in Belfast, questioning him about his voting record on the closure of convent laundries. Sloan was expelled from the Orange Order and helped to found the IOO. A committed evangelical, Sloan labored to make sure that the organization embodied traditional evangelical assumptions. The IOO also presented a working-class challenge to the paternalistic "big house unionism" of the Orange Order and the sitting conservative unionist MPs. It attracted socialists such as Alex Boyd and supported the 1907 Belfast dock strike.[3] The contemporary IOO preserves the evangelical principles of its founding, seeking to promote and defend the Reformed faith. Its membership, after suffering a dip following the dock strike, rebounded and has remained relatively stable at around 1,000, with 45 lodges in Northern Ireland, Scotland, and England.[4] Paisley, though he is not an official member of either the Orange Order or the IOO, regularly speaks at IOO events and is commonly associated with it.

The IOO participates in both cultural and political activities. Its most obvious cultural activity is parading. Other cultural activities are centered around its Orange halls, which are used for a "wide range of activity from Religious Services to Dance Classes" (http://www. ilol.org, accessed July 20, 2003). The IOO has recently committed itself to expanding the use of its halls in an effort to promote "community development." (Interview June 8, 2004). Its renewed emphasis on community development was part of a wider "revisioning" process that involved hiring a consultant to help it make the best use of its assets such as the halls, and to generate more support for the IOO. Another outcome of the revisioning was the establishment of a new organization, Friends of the Institution, that will "be there as a support arm to the organization in general.... focusing in parallel on the issues that the main organization is focused on." (Interview June 8, 2004).

The IOO's political activities include issuing press releases and statements on political issues. Prominent members of the IOO often speak on radio or television programs. The IOO Web site includes press releases on issues such as government structures, the parades commission, policing, and security. It is the only organization in the traditional network to have taken a public stance against the Belfast Agreement. Its press releases and spokesmen argue that the agreement is immoral and unjust, and that it undermines democracy as well as the union with Great Britain. It objects to the early release of prisoners and the participation of Sinn Fein in government without the Irish Republican Army decommissioning.

The Evangelical Protestant Society (EPS)

The EPS was an outgrowth of the UK-based National Union of Protestants (NUP), which expanded to Northern Ireland in 1946. The NUP was concerned with issues such as ecumenism and higher criticism of scriptural texts. The leading figure in the NUP was Norman Porter, a Baptist minister.[5] Paisley was also involved in this organization. According to one activist, the NUP soon split over issues of "personalities" and methodology. Paisley departed, and Porter and his allies formed the EPS. The EPS devoted itself to educating evangelicals about Reformed theology and the errors of Catholic theology (Interview June 5, 2003). The EPS advanced its educational program through its journal, the *Ulster Bulwark*, and by providing speakers for church services and missions. Porter was the EPS's first full-time secretary, followed by Seamus Milligan, who held the post for 36 years. Milligan was replaced in 1997 by an Englishman, Ray Pulman, who continued the *Bulwark* and opened an evangelical bookshop in Belfast. The bookshop failed, nearly wiping out the EPS budget, and Pulman was made redundant. In 2001, Wallace Thompson, already a member of EPS's seven-man council, was appointed secretary and editor of the *Bulwark;* he undertook the work voluntarily from his home. The EPS has seen an increase in membership and activity since then, with the number of *Bulwark* subscriptions growing from about 1,900 to 2,200–2,400. In February 2003, Thompson was appointed as a consultant, paid to work on a one-day per week basis. Under Thompson, the organization has taken a more theological than political tack. Indeed, EPS activists were keen to emphasize that it is not a "political" organization; they said it was rather social or cultural.

The Caleb Foundation

The Caleb Foundation was formally launched in 1998 in the head-quarters of the IOO in Ballymoney. Its aims are to address the per-ceived imbalance and misrepresentation of evangelicalism in the media and the educational sector, and to challenge what they perceive as the dominance of ECONI-style evangelicalism in the public sphere. It takes its name from the Old Testament figure Caleb, one of the spies that Moses sent to scout out the Promised Land.

Caleb's core group of founders (Thompson, George Dawson, David McConaghie, and Mervyn Storey) approached the BBC in 1995 on behalf of the IOO to complain about an imbalance in the BBC's religious broadcasting. They argued that input from ministers and laypeople from small evangelical denominations such as the Evangelical Presbyterian Church, the Free Presbyterian Church, and the Elim Pentecostal Church was absent from religious programs such as *Sunday Sequence* and *Thought for the Day*. This first meeting was not encouraging, so they decided that they would be able to operate more effectively if they were independent of the IOO.

Caleb's founders also believed that evangelical views were not rep-resented in school curriculum. They have argued against the teaching of the theory of evolution without also presenting creationism, rais-ing the issue in the media and meeting with education officials. One activist pointed out that Catholic schools (which instruct children in the Catholic faith) receive state funding. He said that state schools, which are de facto Protestant schools, do not adequately instruct chil-dren in the basics of the Protestant faith (Interview June 5, 2003).

Finally, Caleb's founders were concerned that the "authentic" voice of evangelicalism in the public sphere was being hijacked by ECONI, which was growing in prominence. As one activist said:

> ECONI was becoming very prominent in Northern Ireland, and mas-sively funded by the Northern Ireland Office, getting huge backing by the government, being promoted by the government, and were using government funding to produce booklets and papers that we thought were actually undermining the core evangelical position. [February 17, 2003]

Another activist put it this way:

> There's certainly an organization in Northern Ireland that we would feel we...have been brought into existence to oppose. Which is ECONI, which is Evangelical Contribution on Northern Ireland.

Now, it is not evangelical, it has not made a contribution, the only thing that's right in its name is that it's based in Northern Ireland. [June 5, 2003]

Caleb's efforts have yielded some fruits. Representatives of Caleb have been included on BBC radio programs, the smaller denominations have been featured on the Sunday morning services on the BBC, they have secured a spot for a magazine-style Gospel radio program on Sunday evenings during the summer, and the BBC has produced a new set of guidance notes for its broadcasters. Commentaries by members of Caleb have appeared in the local press. Former Elim Pentecostal minister David McConaghie also was appointed to the Civic Forum, even though Caleb does not approve of the ethos and purpose of the Civic Forum. Caleb decided it was better to lobby for an appointment rather than abstain and lose an opportunity to air its views. Although the organization's public profile seemed to drop a few years after its official founding, activists assured me that this was due more to a lack of time on their part than to a lack of demand for the organization from their traditional evangelical constituency.

The Traditional Evangelical Network in Action

The traditional network extends to other groups. The IOO Web site lists a number of organizations in its network: the Orange Order, the Apprentice Boys of Derry, the Imperial Grand Black Chapter of the British Commonwealth, the Ancient and Illustrious Order of Knights of Malta (sixth language), the Bible Preaching Fellowship, the Caleb Foundation, the Trinitarian Bible Society, the EPS, the Protestant Alliance, and the United Protestant Council.[6] The EPS Web site lists organizations such as Our Inheritance Bible Ministries, the European Institute for Protestant Studies, the Protestant Truth Society, the Orange Order, the IOO, and the Trinitarian Bible Society. Other organizations such as the Society for the Promotion of Reformation in Government (SPRING) and Take Heed could be considered part of this network. The EPS has developed collaborative relationships with groups of traditional evangelical clergy in both the Church of Ireland and Presbyterian Church in Ireland. Caleb has strong relationships with clergy in the small evangelical denominations. Through these ministers and through members of the traditional network speaking at church services and meetings, its message is disseminated to people in the pews. Traditional activists sense that they are gathering momentum. One effect of this "demand" has been the growth of the organizations. They cite as evidence the ability of the EPS to hire a part-time secretary and the increase in subscriptions to the *Bulwark*.

The organizations are not rich by any means, but they seem to have the resources to stay around for awhile. Each IOO lodge has its own budget and is supported by member contributions. The EPS is supported by ad hoc member contributions and legacies. They sponsor an annual Thanksgiving service and appeal each spring; they received about £4,000 at that event in 2004.[7] Caleb has very low overhead costs. Unlike the EPS, it does not publish a magazine or maintain a Web site. Its needs are met by ad hoc contributions. It is currently considering applying for charity status and conducting fundraising in America.

Traditional activists have achieved increased access to government. The IOO "maintains channels of communication with all Unionist Constitutional Parties," seeking to "influence them to an Independent position" (http://www.ilol.org, accessed July 20, 2003). The EPS distributes the *Bulwark* to MLAs. Caleb has actively pursued interaction with British government officials, education officials, representatives of the main unionist parties, representatives of the Community Relations Council (CRC), and the secretary of state for Northern Ireland. They described the process of gaining access to government officials as an uphill battle, but one that will eventually be won.

In sum, the activism of the traditional evangelical network has achieved some of its goals. The network has experienced growth and its organizations have maintained or achieved greater financial stability since the Belfast Agreement. Activists reported increased interaction with government officials and increased inclusion in the public sphere. All these factors indicate that the traditional network is participating in the public sphere and has the potential to play a prophetic role. However, their prophetic potential is impacted by *how* they "frame" or justify their activism, and the way in which that allows them to interact in the public sphere. The following explores how they have reframed their sociopolitical projects, and the implications of that for the transformation of conflict.

The Traditional Evangelical Network: Reframing Sociopolitical Projects

Activists may respond to wider changes by reframing their sociopolitical projects. This process may be more important than whether or not organizations achieve specific goals, especially if they are able to change discourses and perceptions in the public sphere. In postconflict settings, the reframing of sociopolitical projects and the new discourses that this process creates may transform previously polarized social relationships.

The traditional evangelical network is reframing its sociopolitical projects, creating surprising new discourses. Although these discourses have been largely unrecognized in the public sphere, they represent at least a partial change in old ways of thinking and in old strategies and goals. For instance, given the association of traditional evangelicalism with Calvinist conceptions of the relationship between church and state, it might have been expected that traditional evangelicals would advocate some sort of return to a Christian, covenantal political order. This is not the case. Rather, traditional evangelicals use language that indicates they have accepted the assumptions embedded within the British government's civil society approach. They talk about pluralism, equal rights, single-identity work, and nondiscrimination. They criticize the civil society approach, but their criticism centers around how the approach has not worked for them. They are troubled because they believe they are excluded from the officially sanctioned public sphere. This point was underlined when they explained their social activism projects. They have no plans to smash the civil society approach and to replace it with a Christian Ulster; instead, they vow to make it work for them. They say that their most important sociopolitical projects are concerned with moral issues such as homosexuality legislation and school curricula. This indicates that they have accepted the new structure of civil society and are prepared to work within its rules. And like the people in the Free Presbyterian congregation, they hold out hope for a revival.

The following uses narrative analysis to explore the reframing process. The data is presented in the form of representative texts. The texts show how the activists justify their abandonment of Calvinist ideals. This has led them to demand a place at the table, arguing that they face discrimination and marginalization. They have changed partially in response to the new order. Like the Free Presbyterian congregation, they say that their most important tasks are promoting moral reform and praying for revival. Even when they speak of their presence or influence within the Democratic Unionist Party (DUP) (Dawson[8] and Storey were elected as DUP MLAs in the November 2003 elections), they speak of how they may be able to influence educational or moral issues. In this way, they have adjusted their activism along the lines followed by American evangelicals.

Abandoning Calvinist Conceptions of Church and State

Covenantal Calvinism advocates the establishment of a Christian state or a state in which the laws are based on Christian principles. However, the traditional evangelicals I talked to insist that their most important tasks are agitating for moral or social reform—not creating a Christian state or even preserving the union. They still might prefer to live within a Christian United Kingdom, but they recognize that this is most likely impossible. The practical abandonment of Calvinism underlines traditional evangelicals' acceptance of sociopolitical change and a willingness to play by the new rules.

This activist, for example, complains that the Belfast Agreement is governed by a "humanist" ethos and laments the loss of a Christian ethos within society. He links the Belfast Agreement's humanism with excessive concessions to homosexuality, rather than to issues surrounding Irish nationalism or Catholicism. When I asked him if he thought achieving a Christian state was a goal, he said:

> Well, I think we probably recognize that it's not ever going to happen. I think we're moving...further and further away from it. And ideally, yes, we'd like to have a state where all views are taken on board but the British people generally historically have been a Protestant Christian people....I think to be realistic about it we're living in a past age now hankering toward that state based on Christian values. So we try and fight a rearguard action and say these things are wrong and to become almost a remnant of God's people in a difficult sort of social situation. I think that it will get worse....We know that Christians are becoming like Gideon's chosen people. A few people to do what's right. And you battle on, and battle on, and people think you're eccentric at best and weird and sinister at worst. [November 29, 2002]

Now instead of using Calvinism to frame their arguments, traditional evangelicals are framing them in terms of the principles laid out for Northern Ireland's officially sanctioned public sphere. For instance, Section 75 of the Northern Ireland Act (1998) is concerned with protecting the equal rights of all communities. Traditional evangelicals have used Section 75 to argue that a proposed Halloween fireworks display on a Sunday evening violates their equal rights. As an activist said:

> I went to the chief executive of [Belfast City Council]...to ask how the plan for the fireworks display on a Sunday evening...met its obligation under Section 75....Because at its time that's going to adversely impact

on one particular religious group. And even also on age grounds because the population that go to church are elderly.... I know from my own experience going to a meeting one Saturday night in Martyrs' Memorial a couple of years back... Halloween was on a Saturday, and it was dreadful, the noise. You couldn't hear the man speaking. It would be like that if the fireworks display goes overhead. So you use the weapons that they have devised against you to your own advantage. [20 May 2004]

This shift in traditional evangelical discourse is similar to the process that took place when American evangelicals adapted to the breakdown of their privileged relationship with power. Just as the USA's evangelical right still desires a Christian America, traditional evangelicals still desire a Christian Ulster. But like their American counterparts, they seem prepared to settle for a place at the table within Northern Ireland's pluralist civil society. Although the use of this discourse means that evangelicals risk presenting a contradictory stance, it also legitimates the social and political changes that have occurred.

Marginalization and Discrimination

Covenantal Calvinism advocates a privileged place for right religion. However, traditional evangelicals are not framing their demands in these terms. Rather, they are reframing their demands in the language of marginalization and discrimination. They say that evangelicals are being discriminated against on moral or religious grounds. Indeed, one of the main reasons that the Caleb Foundation was formed was because its founders felt that evangelicals were being excluded from the public sphere:

Here in Northern Ireland there's so much about equality, about the need for equality, [but] there is no equality in the legislation in Northern Ireland or coming from Europe which reflects the rights of evangelicals. And everybody can hold any other view under the sun and be legitimized but when it comes to evangelicals, when it comes to saying you believe the Bible, you believe that Jesus Christ is the son of God, you believe that a person needs a personal saving faith in Jesus Christ, you are looked upon as a pariah. There's no legislation, there's no protection.... One of the big issues for Caleb was the lack of representation of our views within the media.... We have an institutionalized ecumenism within broadcasting, within the state, within government, within society, and that has always meant that evangelicals are just squeezed and squeezed and squeezed.... We want to have that [evangelical] community feel that its views [aren't] marginalized but are respected. [June 5, 2003]

The adoption of a discourse that emphasizes marginalization and discrimination is similar to the process that took place when American evangelicals adapted to the breakdown of their privileged relationship with power. Just like American evangelicals, they use the language of nondiscrimination and pluralism to justify the case they make for their place at the table. This line of argument indicates their familiarity with and acceptance of the civil society approach. They complain that the equality legislation does not protect them and that the CRC discriminates against them. They criticize the civil society approach for not living up to its promise. This kind of discourse also legitimates the social and political changes that have occurred.

The Failure of Single-Identity and Cross-community Work

Another way traditional evangelicals justified their activism was by claiming that government approaches to single-identity and cross-community work had failed. Generally, traditional evangelicals advocate a single-identity approach to community relations. The CRC describes single-identity work as aiming to "increase confidence within a community so that people are better able to define their identity and needs in relation to others" ("Community Relations: A Brief Guide" n.d.). The traditional evangelical position that their identity is marginalized and that it should be strengthened seems to fit quite well with the CRC notion of building confidence within a community. But according to this activist, the CRC's efforts to do that have been inadequate:

> You take that away from people and...what you're actually doing is setting up bad relations. People really need to understand themselves before they can understand another. And much of what happens in education, and much of what happens in the pulpits of the main denominations today is stripping away people's understanding of themselves. And there's a pretence that there aren't really differences, when in fact there are differences. And if you understand and have confidence in yourself then you can engage with others without rancor.... I was talking to [a man] from the parades commission...and he was...relating that young Protestant males in Northern Ireland currently are underachieving to the same extent in education as black people in England. And that is to do—these are his theories—that's to

do with a lack of understanding of who they are, a lack of confidence, a lack of cultural identity and all of those issues. Which to some extent reinforces the point I was making. . . . If you look at the board of the Community Relations Council, it's . . . not balanced, it's very weighted. It's an ecumenical council if you like, and . . . the core funding goes in that direction. And while they have a culture program and all those types of programs that are supposed to be single-identity work, there's not a recognition, as yet, of the need for structures in the identity work which will address the void and the vacuum which is within the evangelical Protestant community [and] the Protestant community in general, which is leading to that under-achievement we talked about earlier. [January 30, 2003]

Traditional evangelicals feel rebuffed whenever they try to participate in cross-community work. For example, Caleb attempted to apply for CRC funding for conducting a survey of "ecumenical" clergy asking them what they understood the "evangelical position" to be. Caleb then would have written a report detailing those responses and including Caleb's definition of the evangelical position. The aim of the report would have been to educate both evangelicals and nonevangelicals. Traditional evangelicals also express the desire to "dialogue" with others in the public sphere but feel that they are being deliberately shut out. One activist says that whenever one of their organizations applies to the CRC for funding, they sense that the CRC does not believe that they are sincere in their desire to dialogue.

In sum, traditional evangelicals argue that the cross-community focus of the civil society approach is not working because the CRC and others are not doing what they themselves have pledged to do: build up single identities before engaging in cross-community work. They say that even when they try to participate in cross-community work, they are refused. They claim that others in the public sphere treat them with disdain. This causes traditional evangelicals considerable frustration in their efforts to gain recognition and funding. Again, however, they frame their objections in terms that indicate their acceptance of the rules laid out by the civil society approach. This discourse legitimates the social and political changes that have occurred, including institutions such as the CRC.

Focusing on Moral Activism

These activists say that moral rather than political activism is their top priority. They insist that they are not "political" organizations. They equate the political with party politics, or with specific questions around the Northern Ireland conflict. This has parallels with the way people in the Free Presbyterian congregation "adapted" their traditional identities by focusing on moral issues. Traditional evangelicals' commitment to moral activism is reinforced by the way they talked about their relationship with the DUP. It might have been expected that traditional evangelicals would feel they could depend on the DUP to implement a Calvinist or traditional evangelical political agenda. This is not the case. Rather, traditional activists accept that the DUP has committed itself to participating in a pluralist political order and that their best strategy is to try and influence the DUP on moral and social issues.

For instance, this activist talks of the DUP's success in terms of moving traditional evangelicalism from the margins to the mainstream. But his analysis is tempered by the realization that it is practically impossible for the DUP to conform to an evangelical agenda. He describes a meeting between members of the EPS and the DUP that took place at Stormont:

> [Meeting with DUP representatives] raises your profile, too. I mean you're more mainstream and more center stage again....I would be quite positive about [the overall DUP agenda, specifically proposals about devolved government structures]...I think there's a lot of merit in it. And it's probably practical and pragmatic as well as everything else. But some people, indeed [a particular member of the EPS council], believed it was a weakening of the DUP's position. He said the DUP was the Disappointing Unionist Party rather than the Democratic Unionist Party!...It was really quite funny coming out to the car afterwards, [he] says I'm going to see a doctor in the morning. He says, you know...there's something wrong with me. I'm more right wing than the DUP! [May 20, 2004]

Traditional evangelical activists recognize that their influence is limited, even within the DUP. They are confident that the wider DUP can be persuaded to accept most aspects of their moral agenda although the party is working and must work within the pluralist political order. In this respect, traditional evangelicals within the DUP seem to be taking on a role similar to that of the religious right in the Republican Party in the USA. These discourses indicate that traditional evangelicals have adapted to the new political order, legitimating institutions such as the Assembly. Privileging moral activism above "political" questions such as a united Ireland or the Belfast Agreement underlines the extent to which traditional evangelicals have reframed their sociopolitical projects.

Revivalism

Like the Free Presbyterian congregation, activists in the traditional network hold out hope for a revival. The focus on revivalism is less intense amongst these activists than the people in the congregation but is significant nonetheless. Again, the need for revival is linked to secularism and immorality in society, rather than to dissatisfaction with the political process. This activist sees organizations such as the EPS as setting the stage for a revival:

> People are so sick of materialism, and finding that it doesn't give you peace with God.... I think in all of this there's going to come a great work of the Holy Spirit. Reformation always comes before revival. You've got to set the machinery. You've got to get it right. And then the Spirit of God works. That happened in 1859 here.... All denominations benefited [from the 1859 revival]. [February 13, 2003]

The hope of revival is another way that activists give meaning to their social activism. While a Christian state might not be possible, they believe that their work can either generate revival or (perhaps more realistically) result in small legislative victories on moral issues. Their belief that a revival is necessary is linked to what they perceive as immorality in society, rather than to dangers presented by Catholicism or a united Ireland. By focusing on "moral" rather than "political" issues, they reframe their sociopolitical projects and legitimate the new political order. They are using the new system to try and gain a platform for their moral concerns, rather than trying to smash the system altogether.

Mediating Evangelical Organizations and their Activities

Evangelical Contribution on Northern Ireland (ECONI, Now the Centre for Contemporary Christianity in Ireland [CCCI])

ECONI began with an informal meeting amongst a few evangelicals in 1987. They were motivated by the reaction of some evangelical leaders to the Anglo-Irish Agreement of 1985. Following this agreement, about 200,000 Protestants staged a protest at Belfast City Hall. Paisley addressed the rally and linked religion and politics in a way that these evangelicals believed was wrong. As such, ECONI's founding was motivated in part by the same sort of reasons that motivated

Caleb's founding—dissatisfaction at the way other evangelicals were expressing evangelical identity and politics.

After a series of behind-the-scenes meetings involving about 24 evangelicals, in 1988 the group published a booklet: "For God and His Glory Alone." This document was signed by 200 evangelical leaders. It was intended to be a one-off event, but the people who would form the core of the organization soon decided to develop training programs based on the document. In 1989, the Belfast YMCA started a cross-community program that incorporated in a series of lunchtime seminars the ten biblical principles outlined in "For God and His Glory Alone." A partnership with the YMCA developed, and with successful grant applications made to the Community Relations Unit (later the CRC) and the Joseph Rowntree Charitable Trust, the group translated the principles into a series of Action Packs—workbooks that adapted the principles for further study, organization, and action. In 1992 one of the founders of ECONI, David Porter, was hired as cross-community coordinator at the YMCA, with one-third of his time to be devoted to work for ECONI. A feasibility study was conducted in 1993, and ECONI was launched as an organization in 1994. In 1995, with a full-time staff and a regular magazine, *Lion and Lamb*, ECONI became a charitable trust. By 1998, "For God and His Glory Alone" was on its fourth printing, more than 10,000 copies had been distributed, and more than one-third of all Protestant congregations in Northern Ireland had participated in ECONI initiatives. By 2001, ECONI's staff had increased to 14 (including two part-timers), and *Lion and Lamb*, which began as a four-page newsletter with a circulation of 100, had become a substantial quarterly booklet of 20 to 30 pages with a circulation of 3,000. One activist estimates that about 4,000 people have been on ECONI's mailing list at some point over the last decade and that about 450 Protestant and Catholic clergy (about 90 percent of the clergy are Protestant) are currently on the list. The activist said: "[We have reached] probably about 40–45 percent of the Protestant church, local church people in Northern Ireland" (Interview January 10, 2003).

ECONI's activities have included a yearly conference, an ECONI Sunday once a year, a summer school, and behind-the-scenes diplomacy. Its training program consists of three tracks: the Programme for Christian Peacebuilding, Transforming Communities, and working with "key individuals." The Programme for Christian Peacebuilding features courses that deal with culture, history, and religion—courses that challenge traditional conceptions of identity and teach peacebuilding skills. Transforming Communities involves developing ongoing relationships with groups through a prolonged series of interactive learning experiences. Key individuals are clergy and laypeople in specific locations who are committed to participating in peace initiatives or to using ECONI materials in courses. The ECONI staff member charged with organizing clergy activities interacts with 50 to 55 clergy each year, either through individual contact or through facilitating clergy forums in various locations (Interview February 17, 2003). Porter also was a community relations delegate to the Civic Forum.

In April 2005, ECONI became the CCCI. The CCCI continued with many of the same staff, publications, and activities as ECONI, but with a revamped focus reflecting the social and political changes that had occurred on the island. In the first issue of *Lion and Lamb* after the change to CCCI, Porter wrote: "The character of Ireland in 2005 is virtually unrecognisable from that of 30 years ago....While the challenges of building a peaceful and inclusive society remain, the bigger challenge now facing us is the everyday reality of this change" (2005:4). The organization would continue to have an "evangelical ethos" but would "find new avenues of partnership and service across the Christian traditions" (D. Porter 2005:4).

ECONI/CCCI sees itself as having contributed to the wider work of cross-community peacebuilding through its publications, programs, and other behind-the-scenes action. Its move to become the CCCI reflects a belief that ECONI has done its job and that it now needs a wider scope to address the challenges ahead.

Evangelical Alliance (EA)

The Evangelical Alliance (EA) in Northern Ireland is part of the UK-EA, which also has branches in England, Scotland, and Wales. UK-EA is part of the broader World Evangelical Alliance or World Evangelical Fellowship (WEF). This body had its beginnings in 1846, when it was launched by more than 800 evangelicals from ten countries (http://www.worldevangelical.org, accessed September 2, 2004). The WEF was formalized in 1952, and today it provides a structure for more than 200 million evangelicals in 123 countries (http://www.worldevangelical.org, accessed September 2, 2004).[9]

UK-EA opened a Northern Ireland office in the early 1980s. It has grown from an organization run on a volunteer basis by Presbyterian minister Ken McBride to one headed by a fulltime general secretary (former Baptist minister Steve Cave), a public relations officer, a part-time volunteer community worker, and an administrative secretary.

EA is overseen by a larger volunteer council. Organizations, congregations, denominations, and individuals may become members of EA. In Northern Ireland, 37 congregations (130 ministers) and about 70 organizations are members of EA (Interview January 26, 2004).[10] EA distributes the UK-EA magazine and a Northern Ireland newsletter. Unlike UK-EA, the Northern Ireland-EA has a membership that is not declining.

EA has partnered with ECONI and other organizations to sponsor events. Prior to some elections, EA and ECONI have organized citizenship education programs that encourage people in the churches to vote. EA also works to facilitate links among its 70 member organizations and organizes events for young pastors and leaders of parachurch organizations in its "emerging leaders network." In 2001, EA hired a public relations officer and a volunteer community worker. The public relations officer worked mainly within the Northern Ireland Assembly, attempting to develop relationships with MLAs and to influence legislation through lobbying. Past issues have included lottery funding, gender recognition/transsexuality, and human rights. EA also sponsored prayer breakfasts that were attended by Ulster Unionist Party (UUP), DUP, and Alliance MLAs. During the suspension of the Assembly, the public relations officer advised EA members on political issues and helped to organize community development training workshops and other events. The volunteer community worker has been active in developing a Christian Action Network (CAN) centered around the greater Belfast area. The CAN consists of about 40 Christian organizations that are dedicated to social development and work in disadvantaged areas. These organizations (and some congregations) are registered in a database, which facilitates links and cooperation between them.

Zero28

There is some conceptual difficulty in including Zero28 and ikon in the mediating network. This is due to their insistence that they are not really organizations and to the claim by some of them that they have left evangelicalism behind. There are grounds for calling Zero28 and ikon postevangelical networks all their own. However, this would be a small, isolated Belfast-centered network. In addition, many Zero28 and ikon activists remain active in ECONI/CCCI, EA, or other evangelical organizations. Given this continued participation, I included them in the analysis of the mediating network.

Zero28 was launched in January 1999, but it had its start in a students' informal discussion group at Queen's University Belfast in 1996 and 1997. At this time, it focused solely on issues of peace and reconciliation. Its name was taken from the phone code for Northern Ireland, 028, because that was "the one thing that brought Catholics and Protestants together." Its founding was bound up with feelings of disillusionment about the lack of young people's involvement in politics and about the role of the church in society. One activist said that about 200 people attended Zero28's launch night. Its initial activities included public discussions and private meetings with politicians and discussions in pubs about issues such as parading and sectarianism. Some of the activities attracted media attention, such as when Gerry Adams talked with people in an evangelical church.

Beginning in 2003, Zero28 expanded its focus beyond peace and reconciliation. It organized its activities into five strands: peacemaking, social ethics, environment, arts/creativity, and justice/poverty. This change reflected the relatively peaceful situation in Northern Ireland since the Belfast Agreement. Activities on the expanded program included film nights, a meeting with environmentalists about improving the urban environment in Belfast, and poetry readings. Zero28 sees itself as having contributed to the wider work of cross-community peacebuilding. Its move to include other issues in its remit reflects an awareness of the changing social and political situation in Northern Ireland (a move that is paralleled by ECONI's later conversion to CCCI). However, by June 2007, Zero28 had officially disbanded, a main impetus for this being founder Gareth Higgins's emigration to the USA. At the time of writing, there is discussion about how those who identified with the group might continue to address issues in Northern Ireland.

Ikon

Ikon was founded in December 2001. The principal impetus came from Pete Rollins, a doctoral student in philosophy at Queen's. It draws its name from the concept of an icon. As one activist explained it:

> There are two ways to perceive something. You can perceive it as an idol or an icon. Idol...means to get the essence of a thing...to see it in its absolute essence. And so whenever in the Bible it talks about creating idols, they're creating these graven images of God. So in a sense we're saying well, a lot of Enlightenment religion is about making conceptual idols of God....Domesticating Him via this

discourse....And the other way of perceiving that is an icon. And an icon is that thing that seduces your gaze beyond the visible. There's an idol and it stops your gaze. The icon brings your gaze into another realm. [March 13, 2004]

Ikon's main event has been its monthly meeting, first in the Menagerie Bar and now in the Black Box. Participants are hesitant to call these meetings "services"; one described them as "performance art," another as "transformance art" (art that transforms the viewer or listener). The meetings usually include music, the raising of questions, and a ritualistic element. Past themes have been apocalypse, heresy, Judas, Pilate, desert, a tribute to Johnny Cash, atheism/theism, and champagne and chocolate cake. Attendance varies from 15 to 20 to a capacity crowd of up to 100 people. It is hoped that other people in the bar will observe or become involved in the meeting, something that happened in the jam-packed auditorium on Johnny Cash night. Like Zero28, ikon also has attracted the interest of the media. It has been featured on the popular Sunday morning radio program *Sunday Sequence*. With the publication of Rollins's book in 2006, *How (Not) to Speak of God*, ikon has received increasing attention from evangelical and "emerging church" networks in North America and the United Kingdom. For example, Rollins is quoted in a 2007 *Christianity Today* article as a leading figure in the international emerging church movement (McKnight 2007:36).

Other ikon events are the Last Supper and the Evangelism Project. Ikon's Web site describes the Last Supper as

> an underground, late night, discussion-based community comprised of twelve individuals. At each gathering we invite an influential public figure with controversial views to join us for a meal and some wine in a low-lit, secret location. The evening begins with a brief introduction from our guest in which they inform us about their heartfelt convictions. After this we begin probing these views to ascertain whether they are persuasive. If not, this could well turn out to be...their last supper. (http://www.ikon.org.uk, accessed June 4, 2004)

Previous guests at the Last Supper have included Tony Campolo, Bishop Pat Buckley (a rogue Catholic priest and gay activist), fundamentalist writer Cecil Andrews, and David Ervine of the Progressive Unionist Party (the political wing of the paramilitary Ulster Volunteer Force).

The Evangelism Project is an initiative in which people from ikon encourage other people to evangelize them. They have sought out meetings with groups such as the Russian Orthodox Church, Quakers, Muslims, the Belfast Humanist Society, and the Free Presbyterian Church (including a private meeting with Paisley). Ikon also has a project called Witness, through which they organized candlelight vigils in Belfast for the people who were suffering because of the Iraq war. Ikon sees itself as offering a spiritual alternative to various forms of "Enlightenment religion," including evangelicalism. It provides a space where people can explore spirituality in ways that are not available to them within traditional or mediating evangelicalism.

The Mediating Evangelical
Network in Action

The mediating evangelical network extends to a number of other groups. EA, with its 70 member organizations, sees one of its main purposes as facilitating links between other organizations. ECONI collaborates with Christian groups such as EA, Christian Action Research and Education (CARE), the Corrymeela community, Restoration Ministries, YMCA, the Irish School of Ecumenics, Mediation Northern Ireland, the Raphoe Reconciliation Project, and the Irish Churches Peace and Education Programme. Zero28 and ikon have links with the Catholic Worker Movement, the One Small Step Campaign, the Healing Through Remembering project, Thinkbucket, Youth With A Mission (YWAM), the Clonard Monastery-Fitzroy Presbyterian fellowship group, and City Church. This list is by no means exhaustive. EA and ECONI also have education and support programs designed for clergy.

ECONI and EA have access to much more funding than organizations in the traditional network. ECONI's budget and its number of paid staff dwarf that of smaller organizations such as Caleb and EPS. Its 2003–2004 annual income was £398,276, with £180,551 from grant-making bodies, £161,343 from trusts, £32,131 from voluntary donations, and £24,251 from revenue and gift aid (ECONI Annual Report 2003–2004).[11] Unlike organizations in the traditional network, ECONI receives both core funding and grants from the CRC.

However, now that Northern Ireland has moved into a "postconflict" phase, it is likely that ECONI/CCCI will not be able to maintain such a high level of funding from outside sources. The international "peace pennies" are being diverted to other conflicts around the globe. EA is funded chiefly through the contributions of its members. Although per capita giving amongst EA members in Northern Ireland is higher than in the rest of the UK, EA-Northern Ireland still relies on the central London office to remain solvent. Zero28 and ikon survive in a much more ad hoc way, similar to that of Caleb and the EPS. However, in 2003, Zero28 received a substantial donation from American Tony Campolo's Evangelical Association for the Promotion of Education (EAPE). That funding allowed Zero28 to employ a full-time consultant for nearly a year. Ikon continues on small donations from participants, churches, and the Northern Ireland branch of the international Christian organization YWAM.

EA and ECONI have achieved considerable access to government (this is not a goal of ikon). EA cultivates access through its public relations officer. ECONI has built on the relationships it began to develop when they invited politicians to participate in its Christian Citizenship Forums. Zero28 initially saw one of its goals as being a lobby group, and it has met with members of Sinn Fein, Social Democratic and Labour Party (SDLP), UUP, DUP, Progressive Unionist Party (PUP), EA, the Women's Coalition, and the Green Party.

Although neither EA nor ECONI officially endorsed the Belfast Agreement (Zero28 and ikon had not been formed at the time), members of the organizations participated as individuals in the Yes Campaign during the run-up to the referendum. ECONI even produced "A Time to Decide: Responding Biblically to the Agreement," a document that—while not presenting the agreement as the solution to all of Northern Ireland's problems—nonetheless was overwhelmingly positive (http://www.econi.org/PubVoice/programme. community.decide.0498.html accessed May 1, 2005). ECONI wrote briefings to government officials explaining how they were getting the message wrong and failing to sell the agreement to the unionist community. Some ECONI literature has criticized the work of Christians Against the Agreement, a group that was active at the time of the referendum (Thomson 2002).

Unlike activists in the traditional network, mediating activists do not report the difficulties in interacting with government officials. There are hints that ECONI, in particular, enjoys a degree of influence at this level of high politics that is shut off from organizations in the traditional network. This, however, is likely due to the immediate interests of the government rather than to the deeper beliefs of government ministers. Mediating evangelicals are adopting strategies similar to those used by Canadian evangelicals, who have cultivated discrete relationships with government officials.

Again, unlike organizations in the traditional network, mediating organizations enjoy positive relationships with the media. Beginning with the publicity that ECONI received in 1993 around ECONI

Sunday, media began to approach ECONI representatives to go on television or radio. As one activist described ECONI's developing relationship with the media:

> I was asked on Talkback, a local controversial radio program, "so do you think John Hume's a peacemaker?" Because at that time the Protestant community were giving John Hume stink for talking to Gerry Adams. And I said yes. And the interviewer couldn't believe it...he nearly fell off his chair. Because nearly every Protestant church- man or politician he brought in would sort of say, well John Hume is compromising, supping with the devil type stuff. And I said...I think John Hume, the integrity of his position is trying to reach out and make peace....And that made the media terribly friendly to us. [January 10, 2003]

In sum, the mediating network has grown and its organizations have maintained or achieved financial stability. ECONI/CCCI's growth from a small group of concerned individuals in 1987 to the large organization it is today is perhaps the most dramatic example. EA, ECONI, and Zero28 reported no significant difficulties gaining access to government officials or political parties, nor did they feel they were deliberately excluded from the public sphere. Indeed, ECONI seems to be a model of the kind of organization that is pro- moted in the officially sanctioned public sphere. For instance, ECONI has sought to change traditional evangelical identity (rather than strengthen it) and has promoted cross-community work focused on forgiveness and reconciliation. This also has been the aim of some projects promoted by EA and Zero28, although they do not receive government funding. All these factors indicate that the mediating network is participating fully in the public sphere and has the poten- tial to play a prophetic role. Their prophetic potential is impacted by *how* they frame their activism, and the way in which that allows them to interact in the public sphere. The following explores how they have reframed their sociopolitical projects, and the implications of that for the transformation of conflict.

The Mediating Evangelical Network: Reframing Sociopolitical Projects

For those who equate evangelicalism with Paisleyism, the way the mediating network has adapted to change is astonishing. The discourses they put forward are a total change from the traditional Calvinist dis- courses that usually have been associated with evangelicalism in the public sphere. Mediating evangelicals routinely critiqued Calvinist conceptions of church and state. They saw pluralism as a part of God's divine order that should be celebrated. They drew on Anabaptist conceptions of the church as a model or a community. This language fits nicely with the norms laid out by the British government's civil society approach and has the effect of not only accepting but also promoting that approach. Mediating evangelicals' criticisms are not of the civil society approach, but of apathetic churches' failure to engage with what they think are the more important issues of peacebuilding and social justice. They have embraced the logic of the approach and are working within its rules. Their hope for society is that they can encourage the churches to do likewise.

The following uses narrative analysis to explore how mediating evangelicals are reframing sociopolitical projects. The texts demonstrate how mediating evangelicals critique Calvinism and apathy. They also show that mediating evangelicals see government's civil society approach as an aid to their agenda. For instance, they have been encouraged that single-identity and cross-community activities seem to be working, even if it is at a slower rate than they would like. Mediating evangelicals identify their most important tasks as drawing the attention of the church to peacebuilding and social justice issues and calling on the churches to model good relationships. They do not think that the new frameworks established for Northern Irish civil society hinder them in these tasks. Their discourses are legitimating social and political change, including an enthusiastic embrace of cross-community peacebuilding and pluralism. Their enthusiasm for pluralism and their cooperative relationships with government and other civil society groups have parallels with the discourses and strategies used by Canadian evangelicals. This allows them to participate fully in the public sphere and has gone some way toward changing perceptions of evangelicalism within it.

Critique of Calvinist Models of Church-State Relationships

Anabaptism advocates the separation of church and state. This implies the acceptance of religious and cultural pluralism. In the case of mediating evangelicals in Northern Ireland, achieving something along the lines of the Anabaptist ideal requires critiquing (and then correcting) what they perceive to be the theological errors behind Calvinist conceptions of church-state relationships. They believe that the historical, covenantal relationship between evangelicalism and power in Northern Ireland was wrong and that it contributed to conflict. Former ECONI research officer Alwyn Thomson's *Fields of Vision* (2002) articulates this critique and proposes an Anabaptist-influenced model of church-state-society relations. So, like traditional evangelicals, mediating evangelicals are playing by the new, pluralist civil society rules. But mediating evangelicals see pluralism and a loss of privilege for religion as an asset to the church, rather than an unfortunate reality that requires compromise of the covenantal ideal. This is an important distinction: mediating evangelicals see the Anabaptist model as *right*, while traditional evangelicals are more likely to see it as *necessary*. An ECONI activist put it this way:

> We are in favor of recognizing plurality in society and recognizing that it's simply not good enough for churches to assume that theirs is the only voice in the place and what they say has to shape the way society functions.... By all means [churches should] find ways to make their voice heard in the public square, but they have to realize that it's one voice among many and they are not going to get all they want. And [churches should] realize that debates in the public square and in more secular society are about compromise and pragmatism and therefore we can't go with some idealistic notion of building a Christian society. It's hard for some churches to come to terms with that.... [But] I think it's good within the church to speak a language that ensures that our social and political comment actually is grounded in theological principles. Otherwise you do just become another lobby group without not that much distinctive to say. The church needs to maintain a sense of the distinctiveness of the rationale for what it says even if what it says is the same as what other people are saying. And outside the church, what needs to happen is for government or civil society or whoever's making decisions on these issues...to recognize the church as an important institution in society but not to give it any privileged position. I think we're wrong to ask for that or expect that. [November 29, 2002]

The discourse of anti-Calvinism and pro-pluralism is similar to that used by Canadian evangelicals. Activists argue that in order for church to influence society in the right way, it must not seek a privileged relationship with power. Like traditional evangelicals, they want a place at the table. But they also want evangelicals to realize and admit that they were wrong about seeking to impose a Calvinist model of church-state relations on society. This has the effect of legitimating the social and political changes that have occurred. Repenting—or saying that evangelicals were wrong—is also a way of reaching out to Catholics in the hope of building better relationships. This has the potential to contribute to the transformation of conflict. However, it is not clear that mediating evangelicals' emphasis on repentance speaks for all the evangelicals in Northern Ireland, which lessens its potential impact.

Critique of Apathy

Mediating evangelicals express frustration that other evangelicals are not concerned about social and political issues in Northern Ireland. They believe that these evangelicals do not think that the churches bear any responsibility for contributing to the conflict; that they themselves are not sectarian; and that peace, reconciliation, and forgiveness are not important parts of the churches' mission. One EA activist says she is routinely disappointed by the apathy she encounters when speaking at churches or to other Christian organizations:

> I went to speak at a certain Christian Union at one of the universities and was really disappointed actually in how disinterested they were in politics. You know, a big room with hundreds of young people in it and I felt I was preaching before the real minister started. I was there talking about the elections and they couldn't really have been less interested. It just was not relevant to them. . . . I don't know if it's that in their churches they haven't been taught to relate politics with faith. Maybe they haven't been taught to see the implications that political decision-making has on their lives as young Christians or on their lives full stop. So maybe it's lack of church teaching. Or maybe it's their affluence enables them [to think that] politics is a luxury not a necessity. [January 26, 2004]

These activists echoed the concerns of the participants from the urban Presbyterian congregation who thought that the church was not doing enough to address peace issues and social problems in Northern Ireland. Accordingly, their activism had a sense of urgency—they felt it was up to them to bring these issues to the attention of the churches. This has the effect not only of raising peace issues in the public sphere, but of also legitimating peace as a valid concern for the church. To the extent that they convince churches to engage with these issues, they may contribute to the transformation of conflict.

The Effectiveness of Single-Identity and Cross-Community Work

Mediating evangelicals also justify their activism by citing cases wherein single-identity and cross-community activities have been effective. They talk about effectiveness in terms of observing people change mindsets. Like traditional evangelicals, they stress that people should not be pushed into cross-community work until they are ready. However, the key difference is that while traditional evangelicals want to strengthen traditional identity, mediating evangelicals want to change it. It is this desire to change that makes ECONI's work so favorable to the CRC, which wants its single-identity work to challenge stereotypes. This ECONI activist tells a story about the feedback she received from a woman whose identity and conceptions of what the church should be doing in society had changed as the result of her participation in an ECONI course:

> I'm thinking of one woman who came to one of our courses a couple of years ago. I met her again at a conference...and she came up to me and was very aggressive and said, "you've ruined my life." And that was kind of scary! I was like, "oh, you'd better tell me more." And she talked about how—she'd come on a Journey in Understanding course...[that looked at] the influence of history, religion, culture, and politics, on Christian identity. And the last one was looking at peacemaking....And she said, "Well, I don't fit in anymore." And the disorientation that came having gone through the ideas...when she went back to her church...[she could] no longer settle for the old dispensation....You bring people to the brink of a new experience and you take them into it, and actually the reality is that some things might never be the same again....When somebody comes on our course we're actually much more up front about saying, this might change everything. Not that you're going to rush out and actually hug the next Catholic you see—because sometimes people see that as the natural outworkings of it—it's more, you may look at your church and find that you just can't go along with things. And then it's do you have the courage to actually challenge what you feel needs challenged? [January 10, 2003]

Zero28 and ikon activists have a slightly different approach to single-identity and cross-community work than ECONI or EA. They are not really concerned about building up Protestant identity or even being self-conscious about cross-community work. Rather, they think that focusing on "Protestant" and "Catholic" is missing the point. For instance, in one interview I asked a Zero28 activist, "There's cross-community contact at the events you have, but could you put percentages on Catholics and Protestants?" He said:

> No, and I would never want to. Actually, this is an issue that arose at the very, very start. [We asked do] we need to have a 50/50 Protestant-Catholic management team [on Zero28]? And I can understand the

reasons why people suggested that, but I also have a suspicion that it proceeds from a fairly simplistic understanding of Northern Ireland's problems....So the reason I don't want to answer the question about percentages is that...the very existence of the question implies that there's only two types of people in Northern Ireland and they're the only types of people really worth working with. That nobody else really matters. And the fact that I don't think that those two types of people really exist to start with...that's a problem. And the second thing is that it alienates and excludes me because I can't be easily categorized. [March 10, 2004]

An ikon activist makes a similar point:

I have a friend who's doing a Ph.D. and his whole Ph.D. was on how to bring Protestants and Catholics together under a shared theology, a shared superstructure. I thought, that's the evangelicals' dream, but in one sense I think it's kind of ridiculous. Yes, there is something that joins together...but it's probably as abstract as a shared love for God, the memory of Christ, something like that. But concretely maybe we have our irreconcilable differences. And he spent so much time trying to find out a way of how Protestants and Catholics can have communion together [and I thought]...why can we not just celebrate the difference? [March 13, 2004]

This does not imply that Zero28 and ikon activists disapprove of self-conscious single-identity or cross-community activities on the part of groups such as ECONI or EA. Rather, they are just frustrated that people in Northern Irish society are forced to think in terms of "Protestant" or "Catholic" at all. ECONI and EA seem more willing to accept that this is the way people think, and to operate within that framework. However, ECONI's recent decision to adopt the name Centre for Contemporary *Christianity* in *Ireland* is a symbolic attempt to move beyond Catholic, Protestant, or evangelical categories in Northern Ireland into a more overarching "Christian" category that encompasses the entire island, north and south.[12]

In sum, these activists have accepted the logic of the civil society approach, including its emphasis on cross-community work and the celebration of pluralism. The narratives from the ECONI activists point to instances where single-identity work is bringing about change—even if that change is slow, painful, and difficult. They are not so much criticizing the approach as observing that it takes a very long time. These narratives also capture their enthusiasm for pluralism, as encapsulated by the ikon activist's desire to "celebrate the difference." This echoes the line often taken by Canadian evangelicals, who tend to view pluralism in a positive light. These discourses legitimate the government's civil society approach, providing theological justifications for pluralism and cross-community cooperation.

Focusing on Social Justice

Mediating evangelicals say their activism is motivated by the churches' failure to address what they call "social justice" issues. These range from issues particular to Northern Ireland, such as peacebuilding and forgiveness in politics, to broader, international issues such as fair trade and the environment. Mediating evangelicals are articulating a twin-track strategy: concentrating on peacebuilding in Northern Ireland while raising wider social issues that evangelical churches have not focused upon.

For instance, ECONI and Zero28 both started as organizations dedicated to peacebuilding in Northern Ireland, in part because they were concerned that the churches were not doing enough in this area. In recent years, both have expanded their mandate to wider issues. This is reflected in ECONI's decision to become CCCI, with its emphases expanding to "conflict," "community," and "citizenship" ("Biblical Faith for a Changing World" 2005:9). Two recent editions of *Lion and Lamb* focused on multiculturalism in churches in Northern Ireland and the Republic of Ireland and on racism.[13] The racism edition addressed the issue in both Northern Ireland and the Republic and generated such a response that it exhausted its print run of 3,000 copies (Rankin 2005:1). ECONI also has had workshops and conferences on the environment and women in the church. In 2003, Zero28 developed its five strands: peacemaking, social ethics, environment, arts/creativity, and justice/poverty. EA's development of the CAN reflects a greater concern with social issues, as does ikon's Witness project. This EA activist says that working for social justice is just as important a part of the Gospel as getting people "saved." He says EA's CAN and other projects connect the Gospel with social justice:

> The work of having a Christian Action Network or...encouraging the church to make a presence at street level [is] not just to be out to get people saved. I believe [social justice] is a Gospel imperative. I don't believe it's secondary to the Gospel, I believe it's part of the Gospel. You can't have scissors with one blade....And I don't believe the Gospel we offer is conditional. We don't say...you can have our soup if you accept my tract, I think we need to be more generous than that. So the concept of being out there is critical for me. And I've been

involved in supporting a project...which was trying to bring a presence at street level in an area where there were 13 churches, all doing their own thing, but not necessarily having much impact...in an urban deprived area. And over a number of years I have with others been pleased to see those churches beginning to do things at street level and not just in their church buildings. A café. Support for mothers, support for children, after school clubs...things like that....Challenging churches to be involved increasingly in the community is part of what I believe we need alongside our theological and other convictions. [March 10, 2004]

Mediating evangelicals recognize that their influence within the churches is limited—they can not seem to get the churches to concentrate on the issues they think are important. Like traditional evangelicals, they are trying to draw attention to the issues that they think are important. Traditional evangelicals are trying to draw attention to "moral" issues by targeting the wider society, while mediating evangelicals are trying to draw attention to "social justice" issues by targeting the churches. Both the moral and the social justice discourses represent an expansion of evangelical sociopolitical projects beyond the issues of Protestant (ethnonational) identity and preventing a united Ireland, with which evangelicals are so often identified.

The Church as a Model Community

Finally, mediating evangelicals are adamant about their desire for the church to do what they think it *should* be doing. They do not think the church should be working toward a revival, like the traditional evangelicals. Rather, they talk about the church "being" the church— acting as a model for society or being a community. They think it is their job, as organizations, to urge the church to do its proper job. This Anabaptist-influenced idea of the church draws on the theology of Hauerwas and Yoder. Some activists conceive their role in monastic terms, recalling how historically it is often the theology and prayer of the people in monasteries that draws the wider church back to its true mission. Chapter six of Thomson's *Fields of Vision* (2002) makes a theological argument for a "model" interpretation of the church.

An integral part of the model is living peaceably with difference. That means developing relationships in which difference or disagreement does not mean that people stop interacting with one another. Mediating evangelicals say that Catholics and Protestants in the wider society are too often separated, and that their interaction is too often marked by suspicion, fear, indifference, or violence. As this ECONI activist said:

> Part of our strategy through churches is about getting churches to look at the nature of their relationships with others in the community, whether it's people from other Protestant traditions or whether it's people from the Catholic tradition. And [to look at] how they can actually begin to make contacts based on these relationships. Through some of the public meetings we've done, like bringing in Sinn Fein and the SDLP, it's about setting up a conversation that moves beyond simply accepting that some people vote for them. [It's about accepting that] they have to have some role. One thing we've never agreed with is the position among some people in the ecumenically-inclined churches that you need to create some kind of ecumenical reconciliation between churches before you can create reconciliation in society....Partly because it's wrong, partly because...it gives churches greater significance than they actually have....One of the things we need to learn in this community is not to sink our differences but to find ways to have our differences constructively and in a way that has some measure of relationship, where there is still difference. And actually churches instead of pretending those things aren't there can actually do a good job of modeling how you can relate them constructively in relationships within that. [December 29, 2002]

Rather than hoping for a revival, mediating evangelicals hope to make the church a better community. This community would be a place where people lived peaceably with their differences without necessarily trying to change each other. It was hoped that the wider society would observe the church modeling good relationships and would follow suit. This theological conception of the church has also been prominent within Canadian evangelicalism, both historically and in the justifications for activism supplied by contemporary organizations such as the Evangelical Fellowship of Canada.

Conclusions

This chapter analyzed how networks of traditional and mediating evangelicals are adjusting to social and political change. It focused on what networks of organizations are actually doing, and how they have reframed their sociopolitical projects. This demonstrated that processes of change are taking place (albeit slowly), and that these processes have implications for the transformation of conflict.

Activists in the traditional network were not entirely comfortable with the changes that have occurred, especially with the Belfast

Agreement and the way it has been implemented. They felt that they were disparaged or deliberately excluded from the public sphere. However, their activism had resulted in growth and financial stability for their organizations, increased interaction with government officials, and increased inclusion in the public sphere (particularly through the media). They felt confident that they would be able to broker for more influence in the DUP, especially on moral issues, when the Assembly was restored. However, they reframed their activism in terms that signified at least a partial change in traditional evangelical ideals. For practical purposes, they had abandoned the covenantal Calvinist ideal about the relationship between church and state. Rather than arguing for a privileged place for "right religion," they made their case for their inclusion in the public sphere on the grounds that they were facing marginalization and discrimination. They criticized single-identity and cross-community work because it was not working for them. This discourse signaled their willingness to work within the new civil society structures. It legitimated civil society's pluralist ethos, including institutions such as the CRC. Their decision to reframe their sociopolitical projects in terms of moral issues (and to hope for revival) also, in a more indirect way, confirmed their acceptance of the current context. Stopping abortions seemed more important than stopping a united Ireland. They were able to participate at least partially in the public sphere, even if they complained that they were frequently misunderstood and misrepresented.

Activists in the mediating network were enthusiastic about the changes that had occurred. When they criticized the changes, it was because they had not gone far enough or were moving at too slow a pace. They did not necessarily feel excluded from the public sphere but believed that the churches excluded themselves from public debate through apathy and indecision. Their activism resulted in growth and financial stability for the organizations. In the case of ECONI/CCCI, this growth was quite spectacular and involved impressive competence in gaining government funding. Because their agenda currently dovetails with the goals of government's civil society approach, they are accepted and even promoted by the state. This is ironic, given that Anabaptist theology rejects sociopolitical power. Mediating activists must deal with the tensions between their theological beliefs and the practical ways they are interacting with the government. They say they must remain vigilant lest their work be co-opted or restrained for state purposes. But the extent that they are dependent on government funds (especially ECONI/CCCI) forces them to maintain a delicate balance.

In addition, mediating activists had abandoned covenantal Calvinist assumptions about the relationship between church and state, not because they were no longer realistic, but because they thought they were wrong. They urged evangelicals to repent for the way that evangelicalism had been bound up with power in Northern Ireland, believing that this was necessary for forgiveness and reconciliation in the wider society. When they criticized single-identity and cross-community work, they did not say that it was not working; they said that it was not working quickly enough. This discourse signaled their enthusiasm for the new civil society structures. This allowed them to participate fully in the officially sanctioned public sphere and had gone some way toward changing wider perceptions of evangelicalism. It also legitimated civil society's pluralist ethos, including institutions such as the CRC. Finally, expanding their focus from peacebuilding to local and global "social justice" issues signaled that they no longer thought it necessary to focus exclusively on peacebuilding in Northern Ireland. In an indirect way, this also confirmed their acceptance of the current context.

Traditional and mediating evangelicals viewed each other with a great deal of suspicion and disapproval. They clearly have different visions of what society should be like, based on very different interpretations of the Bible. But even though this disagreement remains, it is significant that neither group of evangelicals has outright rejected all of the principles of the Belfast Agreement. Even traditional evangelicals have displayed a capacity to adapt and to change.

Chapter 6

Conclusions

The orthodox interpretation of Northern Irish evangelicalism equates Paisleyism with evangelicalism. Rev. Ian Paisley's critics keep a running tally of his "sins," which range from blanket accusations that he was responsible for starting the Troubles, that he has used religious rhetoric to incite loyalist paramilitaries to violence, and that he leads a political party that refuses to share power with Catholics.[1] All evangelicals have been judged guilty by association. This has meant that evangelicals have been considered barriers to the Northern Ireland peace process, so much so that other secular and religious actors have argued for their exclusion from the public sphere.

This research calls for a reassessment of this received and often unquestioned orthodoxy. This is important not only because the accepted orthodoxy is wrong, but also because it results in a failure to understand evangelicalism's contribution to the transformation of the conflict. Ignoring the evangelical aspect of the conflict will not make it go away. Given evangelicalism's historical importance and continuing social significance within the Protestant community, engaging with it is a necessary—and not just optional—step in the process of conflict transformation.

Recent research has begun to challenge the orthodox interpretation of evangelicalism (Mitchel 2003; Mitchell 2003; Jordan 2001; Brewer and Higgins 1998; Thomson 2002; F. Porter 2002). This research builds on that and takes it a step further. It makes new distinctions between the types of evangelicalism that are emerging and provides insight into the implications of those changes. As such, the findings and conclusions presented here are sometimes startling and unexpected. For those who have willfully ignored evangelicalism, the existence of evangelicals who repudiate Paisley and his religious and political projects is something of a revelation. For those who are aware of these other expressions of evangelicalism, the nature of the changes on the part of evangelicals who are sympathetic to Paisley is equally surprising.

By showing the processes of identity change and the reframing of sociopolitical projects at work, this research lends support to the argument that the Northern Ireland conflict can be transformed—not just managed with varying degrees of effectiveness. Analysts such as McGarry and O'Leary have doubted that this could ever be the case, dismissing the contributions of "transformers" as "utopian and mistaken analyses" (2004:18). Similarly, Tonge has claimed that "the transformationist approach fails to accept that ethnic divisions exist because humans develop a sense of identity that they wish to retain" and that "transformationists appear to think that holding an identity and a concomitant political position is somehow retarded and should be overcome" (2005:256–257). These criticisms are misguided. This research has demonstrated that evangelicals are retaining and modifying their identities—not "overcoming" them—and that this process is already contributing to the transformation of conflict. Moreover, it is a process that can occur even if political institutions are not up and running.

Evangelical Diversity

Popular perceptions of Northern Irish evangelicalism have been beset by the myth of an "ideal type" evangelical, often embodied in the figure of Paisley. However, the perpetuation of the Paisley stereotype masks important differences within evangelicalism, precludes the possibility of changes within evangelicalism, and fails to draw out the implications of those changes for the possible transformation of conflict.

Other recent works of research have gone some way toward categorizing the types of evangelicalism or evangelical identities (Mitchell 2006:117–132; Jordan 2001; Mitchel 2003; Brewer and Higgins 1998). My research built on these earlier efforts, collapsing the identities into four empirical types: traditional, mediating, pietist, and postevangelical. I based these empirical types on the beliefs that are most important in divided societies: beliefs about the proper relationship between church and state, about religious or cultural pluralism, and about violence and peace. These empirical types provided a framework for understanding what evangelicalism is like "on the ground," and for charting people's religious identity change.

Traditional evangelicals usually are associated (if not equated) with Paisleyism. They have Calvinist-informed beliefs about a covenantal relationship between church and state, a privileged place for right religion, and the selective use of violence as a last resort. Mediating evangelicals present the clearest and most prominent challenge to traditional evangelicalism and indeed emerged in Northern Ireland as part of a concerted attempt to challenge the religious (and political) beliefs of traditional evangelicals. They have Anabaptist-informed beliefs about the separation of church and state and are enthusiastic about pluralism and nonviolence. Postevangelicals present a challenge not only to traditional but also to mediating evangelicalism. Like mediating evangelicals, they favor a strict separation of church and state, religious and cultural pluralism, and nonviolence. But their enthusiasm for pluralism goes above and beyond that of mediating evangelicals, incorporating other Christian and religious traditions and more radical ways of interpreting the Bible. Pietist evangelicals simply advocate withdrawal from society and politics.

Analyses of the conflict and the settlement have often failed to take account of this evangelical diversity. Analyses of mediating evangelicalism have been sparse; a chapter on Evangelical Contribution on Northern Ireland (ECONI) in Mitchel's work is the most extended treatment of their impact thus far. Postevangelicals have not entered into academic analyses at all. In addition, the stability of the beliefs of traditional evangelicals have been too much taken for granted, with very little development from the major works of Bruce, which date from the mid-1980s. This has meant that conceptions of evangelical identity remain essentialized, with few recognizing the diversity and change within evangelicalism. This is a crucial omission, because in conflicts in which religion is bound up with identity, religious change is an important part of the process whereby identities may loosen and become less polarized.

The Significance of Evangelical Diversity and Change

The emergence of diverse evangelical identities and different forms of evangelical activism is a complex story. In the few previous works that have recognized and analyzed evangelical diversity, a "new orthodoxy" is already starting to emerge. This orthodoxy focuses on the work of (mediating evangelical) groups such as ECONI, praising their contribution to the changing of oppositional identities and the steps they have taken toward transforming the conflict. Traditional evangelicals are criticized as obstructionists who cannot or will not change. This is an interpretation that bears weight but, if presented too simplistically, runs the risk of creating new evangelical myths. It may overestimate the impact of mediating evangelicals and underestimate the potential of traditional evangelicals.

Appleby (2000) has argued that religious actors are most effective when they critique the role that religion has played in conflict and put forward distinctly religious perspectives on how to resolve it. Mediating evangelicals certainly fit this bill. Mediating evangelicals and their postevangelical allies have had a transformative impact on the Northern Ireland conflict at a number of levels. This was demonstrated in both the congregational and the organizational studies. Within the congregations, that transformative impact was seen in mediating evangelicals' enthusiasm for pluralism and for change. They drew on resources within the mediating identity that allowed them to assimilate to changed social and political circumstances. These resources included an emphasis on peacebuilding, forgiveness and reconciliation, and pluralist (Catholic-Protestant) dialogue. Similarly, postevangelicals remained motivated by their faith to focus on peacebuilding and social justice. A consequence of this was that most mediating and postevangelicals seemed determined to "do something" to bring about change—even if it was only taking "small steps." Depending on the culture of their congregation and their personal resources, they were equipped to drive change in others and in themselves. They also complained that others in the (evangelical) churches were not focusing enough on these issues, especially peacebuilding.

Within the organizations, mediating evangelicalism's transformative impact was demonstrated by its emergence from Paisley's considerable shadow. Some mediating activists said that people in the wider community now recognized the different forms of evangelicalism. The wider project of the mediating network is to change (traditional) evangelical identity, and to convince evangelicals (and the wider Protestant community) to become involved in building a new, pluralist order. They framed their justifications for this project in terms that drew on evangelical history and that recognized Northern Ireland's changed sociostructural conditions. They said that they were convinced that the way that Northern Irish evangelicals had applied Calvinist principles such as the covenant and the chosen people to their own situation throughout their history had been a mistake. Mediating evangelicals believed that this form of evangelicalism had contributed to the conflict in Northern Ireland, arguing that at best it had helped to create boundaries between Protestant and Catholic, and at worst it had incited Protestants to violence. They framed their arguments in terms that called for a rejection of Calvinist conceptions of evangelical identity, replacing them with conceptions drawn from the Anabaptist tradition. These included a conception of the church as a model or a community and a celebration of pluralism and differences as good things that were part of God's divine order. They believed that the position of prestige that evangelicalism had occupied had been wrong, and that the new structural conditions presented them with an unprecedented opportunity to contribute to change in new directions. Mediating evangelicals were enthusiastic about the potential of the Belfast Agreement, British government reforms, and cross-community work within civil society. Their criticisms centered around the apathy of evangelicals and the failure of the churches to embrace these changes and to devote themselves to the issues that mediating evangelicals thought were important. As such, this language dovetailed with the secular language of pluralism, equality, and parity of esteem reflected in the Belfast Agreement and promoted by the British government. It legitimated the Belfast Agreement and the emerging, officially sanctioned public sphere and provided evangelicals with religious and moral justifications for participating in the new order.

However, it is not clear to what extent mediating evangelicalism has penetrated the grass roots. Just as cross-community peace and reconciliation organizations may be unrepresentative of what Northern Irish civil society is actually like, mediating evangelicals may be unrepresentative of what Northern Irish evangelicalism is actually like. Postevangelicals are most assuredly unrepresentative. Even though their emergence in the mid-1980s and their continued participation in the public sphere have been impressive, the depth of their support base is not clear. As such, mediating and postevangelical activism may be slow in producing the desired results of changing identities, constructing new sociopolitical projects, and transforming the conflict.

Also, as mediating and postevangelicals move ever further from traditional evangelicalism, it is possible that they will lose touch with traditional evangelicals. This is already apparent in the strained relations between ECONI/CCCI and groups such as Caleb or the Evangelical Protestant Society (EPS). The further mediating and postevangelicals move from traditional evangelicals, the more difficult it becomes for them to contribute to the changes within traditional evangelicalism that they would like to see. Indeed, as mediating and postevangelicals continue to move, it seems they are abandoning the aspects of their projects that focus on contributing to change within traditional evangelicalism. In his discussion of ECONI, Mitchel recognized the difficulties facing mediating (what he calls "open") evangelicalism:

> A stated aim of ECONI is to be a catalyst for change at a grassroots level through resourcing and influencing Christian leaders. Its highly activist approach derives from this commitment. However, the very nature of an open identity may militate against the achievement of this goal. My review of ECONI activities suggests that participants will be those who are already sympathetic to a critique of religious national-ism.... ECONI adopts an objective approach and a deliberate aban-donment of the emotive tools of nationalism (the use of ritual, symbols, and historical myth). It refuses to indulge in the politics of fear and siege. It displays a willingness to question the sacred cows of closed evangelicalism in a search for truth and a tendency to deal in the subtleties of theological debate. It self-consciously attempts to embrace diversity. All of these characteristics stand in contrast to the simple, popular, unifying and emotional appeal of nationalist religion.... Operating in a context of shrinking unionist support for the [Belfast Agreement] and increasing fear and distrust that fuels unionist myths, the group faces considerable obstacles in its goal of eroding the ideological base of closed evangelicalism. (2003:297–298)

This is a situation in which mediating and postevangelicals are just "preaching to the choir." For example, Mitchel complains that "closed" evangelicals (who he associates with Paisleyism and Orangeism) minimize and dismiss what ECONI and the Presbyterian Church in Ireland have had to say about religion and the Northern Ireland conflict. Mediating evangelicals are troubled by the failure of traditional evangelicals to agree with them, and it is possible that engagement between mediating and traditional evangelicals will break down altogether. Their zeal to "change" traditional evangeli-cals also may lead mediating evangelicals to do that for which they hold traditional evangelicals culpable: refusing to listen.

This makes it easy to overlook significant developments within tra-ditional evangelicalism. The focus on traditional evangelicals' refusal to recognize, criticize, or repent for the role of evangelicalism in the conflict has obscured subtle changes in the way that they are begin-ning to articulate their identities and justify their activism. For instance, traditional evangelicals within the congregations I studied claimed that the core of their religious beliefs had not changed, and this was reflected in their wish to have "Christian" or moral men in government, and "terrorists" out. Significantly, however, this did not automatically lead them to focus on rebuilding a "Protestant parliament for a Protestant people." This was the case for the Free Presbyterian congregation, whose members concentrated their energies on moral activism. This also was the case for traditional evangelicals in the rural Presbyterian congregation, who talked about doing what they could in the local context. It is significant that they have adapted to—and not rejected outright—the new order. While adaptation may not signal a quick and enthusiastic embrace of the new order, it is at least a reluctant shaking of hands with it. As such, traditional evangelicals may not be the obstructionist enemies of the Belfast Agreement that they are often assumed to be. They are changing, albeit slowly and painfully, and their ability to adapt may reinforce similar trends in the wider Protestant community.

This process of adaptation was illustrated even more starkly by the way in which organizations in the traditional network reframed their sociopolitical projects. They did not focus on rebuilding a Protestant parliament for a Protestant people. They did not even oppose the Belfast Agreement as stridently as might be expected. Rather, they refocused their activism on select issues on which they believed they could make an impact. These issues were "social" or "moral," such as education and homosexuality legislation. As such, their goals were relatively modest, and their rhetoric was relatively restrained. They said that a Christian country was unrealistic. They had in effect abandoned traditional Calvinist ideals, even if Calvinist concepts were still reflected in their narratives. They may not have liked some of the changes that have recently occurred in Northern Ireland, but they were working within the pluralist framework. Traditional evangelicals said that their most important tasks were promoting moral reform and praying for revival. Strategically, they aimed to accomplish their goals by using their organizations to make their voices heard in the public sphere and by attempting to barter for influence within the Democratic Unionist Party (DUP). Even when they spoke of the DUP, they talked about how they might be able to influence the party on educational or moral issues. In other words, they were not out to smash the new structures, but they recognized that the structures kept them from achieving the more far-reaching goals they might have had in the past. Their narratives (while still at times drawing on historic evangelical concepts) reflected a degree of practical acceptance and legitimation of the new order.

That said, traditional evangelicals' decision to focus on "moral" issues is unlikely to have a great appeal outside the evangelical constituency. Bruce (1986) and Mitchel (2003) have argued that non-evangelicals vote for the DUP in spite of some of the party's past stands on moral issues and Sunday shop openings. This is because they see the DUP as a more vigilant defender of the union in times of crisis. If politics in Northern Ireland becomes more "normal" or less crisis-ridden, such moral stands may hurt the electoral fortunes of the DUP (Farrington 2006, 2004a). Eventually, traditional evangelicals may find themselves marginalized in their own party if they continue to pursue the moral agenda.[2]

In addition, focusing on traditional evangelicals' refusal to recognize, criticize, or repent for evangelicalism's former relationship with power runs the risk of overlooking the felt experience of traditional evangelicals, especially if they have had direct experience of violence. While it has been argued here that evangelicalism once had a privileged relationship with power in Northern Ireland, this power did not necessarily have a felt impact in the lives of ordinary evangelicals. Indeed, many evangelicals and other Protestants have felt powerless in the face of Irish Republican Army (IRA) violence. They also believe that it is unjust that paramilitary violence seems to have been rewarded with political power. Accusing relatively powerless Protestants of participating in the oppression of Catholics and urging them to "forgive" the IRA often does little to contribute to the healing of wounds that could lead to a transformation of the conflict. Indeed, there was a sense amongst the traditional evangelicals that I talked with that Protestants were expected to do all the repenting and forgiving, whilst the IRA had been exempted from these expectations. This feeling was reflected in Paisley's comments during December 2004 negotiations, when he told supporters at a meeting in Ballymena that the IRA should be "humiliated" and must "repent in sackcloth and ashes." Brewer expressed it in much more moderate terms when he wrote of Protestants that "the simple word 'sorry' would say so much to these people" (2003a:106).

The evangelical diversity analyzed here has implications for the transformation of conflict in Northern Ireland. Brewer (2003b) has written that the transformation of the conflict requires not just personal attitudes of nonsectarianism on the part of religious actors, but also concerted antisectarian programs. This research has demonstrated that mediating and postevangelical activists have been prominent in devising such initiatives and participating in them. Their participation often was linked with changes in their personal evangelical identity. Mediating and postevangelicals testify that some of their efforts have brought about positive change, albeit more slowly than they would like. Following Brewer, the continued activism of mediating and postevangelicals has the potential to contribute to conflict transformation.

Pietist and traditional evangelicals certainly perceived themselves as nonsectarian (despite their critics' assertions to the contrary). Pietists, by definition withdrawn from politics, are not then likely to participate in "antisectarian" efforts. Traditional evangelicals often disapproved of the sort of programs that Brewer would describe as "antisectarian." As such, it would be easy to conclude that pietist and traditional evangelicals have little potential to contribute to conflict transformation. However, this is not entirely accurate. Pietist evangelicals have the potential to be mobilized, especially if they are situated within a congregation in which activism is a part of its culture. And traditional evangelicals, even if they are not participating in antisectarian efforts, have changed to a surprising degree. Recognizing that as well as acknowledging the hurt and concern of those who believe that significant aspects of the implementation of the Belfast Agreement have been unjust are necessary correctives to the emerging myth of "good" mediating evangelicals and "bad" traditional evangelicals.

Finally, the emergence of diverse evangelical sociopolitical projects counters the perception that all Protestant or unionist responses to the wider peace process have been fatalistic or reactionary (N. Porter 2003; Aughey 2001). Farrington (2006) has argued that unionists see the peace process as something that has happened to them, and that "unionists have not exerted much intellectual energy in constructing a vision as to where the process should be going" (2006: 127). However, mediating and postevangelicals perceive themselves as participating in and shaping the wider peace process, and they claim to have constructed a postsectarian vision of society based on forgiveness and reconciliation. As for unionists who feel alienated from the peace process, Farrington points out that while they may use a confrontational style and oppose the Belfast Agreement, their words often actually affirm the principles of the agreement that they oppose. He quotes the DUP's Jeffrey Donaldson who—like many of the traditional evangelicals in this study—stresses the necessity of an "inclusive" society but sees the Belfast Agreement as an ineffective way of achieving that objective. The tendency has been to equate traditional evangelicals' and antiagreement unionists' confrontational *style* with opposition to key principles in the agreement (such as power-sharing), and to accuse them of providing no other political alternative to a Protestant parliament for a Protestant people. The tendency of unionism's critics to complain about *how* they put their case has meant that Protestant responses to the peace process have been misunderstood and their concerns (which may be legitimate) dismissed.

Evangelical Networks: Linking SocioPolitical Projects and Identity Change

Networks are the most dynamic religious structures when it comes to shaping and contributing to change. This research demonstrated clear and direct links amongst traditional organizations and amongst mediating/postevangelical organizations. However, religious networks are not limited to organizations, and they are most effective if they include congregations. The links between the congregations and the networked organizations analyzed here were not always clear and direct, but there was evidence that such links exist and are significant in the process of change.

This was most clearly demonstrated amongst the urban mediating evangelicals and the postevangelicals (who all also lived in an urban area). Without exception, the urban mediating and postevangelicals had participated in networked religious, evangelical, or secular organizations, or in projects sponsored by such organizations. Some linked their participation directly with changes in their identities, including changes in the way they thought about faith and politics. Their participation in the wider mediating/postevangelical network also made them aware that the churches were not "doing enough" in the areas of peacebuilding and social justice. This infused them with a sense of mission to convince others in the churches to take up these causes. Exposure to the sociopolitical projects advocated within the networked organizations contributed to identity change amongst these mediating evangelicals.

This contrasts with the experiences of the mediating evangelicals in the rural Church of Ireland, who rarely mentioned participation in networked organizations. However, they were not entirely isolated from the effects of the mediating/postevangelical network. The rector had participated in ECONI events and had brought ECONI to his parish to deliver a cross-community study on faith and politics. In this diffuse way, the sociopolitical projects of the mediating/postevangelical network reached the parish. Exposure to the sociopolitical projects advocated within the networked organizations was consistently given to them through parish life rather than through the organizations themselves. Although links here between organizations and the members of the congregation were not explicit, the consequences were similar. Participation in their parish had contributed to changes in their religious identities. Moreover, their sense that they should take "small steps" to build peace resembled the urban mediating evangelicals' sense that the churches should "do something."

Links between the traditional evangelicals in the congregations and the traditional organizations were not as clear. In the rural Presbyterian congregation, three of the four people that had traditional identities (including the pastor) were involved with various branches of the loyal orders in some way. Apart from the pastor, who had used Orange platforms to criticize the Belfast Agreement, they did not make specific links between their participation in these organizations and the way they thought about faith and politics, or the way they put their religious beliefs into action. However, they echoed the beliefs of their pastor when they criticized the "immorality" of the Belfast Agreement. In the urban Free Presbyterian congregation, no one mentioned participating in religious or evangelical organizations outside of their congregation. However, their congregation operated like a well-organized interest group, mobilizing people for a range of activities that centered around moral issues. Their pastor, although not a member of any of the loyal orders, Caleb, or the EPS, spoke publicly on moral issues and members of the congregation talked about hearing him on television and the radio. The way the pastor and members of the congregation justified their activism paralleled the way in which activists in the traditional organizations justified their renewed focus on moral issues. This research cannot demonstrate conclusively that exposure to the sociopolitical projects of the traditional organizations contributed to identity change or to a refocusing on moral issues amongst traditional evangelicals in the congregations. That said, it is impossible to overstate the uniformity of the morality-centered narratives of the traditional evangelicals (especially the urban Free Presbyterians), and those of the activists in the traditional organizations. The discourses of the traditional organizations reached traditional evangelicals in the congregations in a diffuse way, mediated through their pastors.

The links identified here indicate that evangelical networks are "working," and in particular that the sociopolitical projects articulated by activists in organizations are reaching and impacting people in the pews. This is a significant finding, especially in the case of the mediating/postevangelical network. For example, ECONI/CCCI has been criticized for being unrepresentative of evangelicals at the grass roots, and for being propped up by the government. While there may be some substance to those criticisms, this research demonstrates that this is not the only story. ECONI/CCCI and its partners *have* contributed to changes in evangelical identities. This research can quantify neither the extent of identity change within evangelicalism nor the extent that mediating organizations have contributed to change amongst individuals and congregations. However, there is survey evidence that mediating/postevangelicals are more numerous than is usually supposed. For instance, Mitchell and Tilley (2004) have isolated a sample of evangelicals from the Social Attitudes Surveys for Northern Ireland that repudiates the common assumption that evangelicalism translates into support for the DUP. This assumption is based on the premise that traditional evangelicalism is the dominant voice of Northern Irish evangelicalism. But their work demonstrates that evangelicals are *less likely* than other Protestants to vote for the DUP. Certainly, only one mediating/postevangelical in my sample admitted to voting for the DUP, and even that he did only once in his 59 years.

Evangelicalism, Identity, and Political Institutions

Those who argue that it is impossible for the Northern Ireland conflict to be transformed advocate an institutional form of conflict management, such as consociationalism (McGarry and O'Leary 2004; Tonge 2005). Consociationalism includes proportional power-sharing and is meant to make sure that the autonomy of competing groups is reaffirmed and protected. It operates under the assumption that attempts to "force" competing groups to change are impossible. O'Leary (1997) was operating under these assumptions when he argued that the British government between 1979 and 1997 was engaged in "slow ethnonational policy learning." He argued that the Sunningdale Agreement of 1973 included the consociational measures necessary for a settlement, but that after its collapse the British government failed to introduce settlements that included the necessary consociational institutions. Accordingly, it was not until the election of the Labour government in 1997 that British government policy changed, and this resulted in the consociational Belfast Agreement.

However, O'Leary's institution-focused account fails to consider that the British government faced structural constraints that made it impossible to impose a Sunningdale-style settlement (Dixon 2001a).[3]

It also fails to consider significant changes that took place between the Sunningdale and Belfast Agreements. These included changes in power relationships between Protestants and Catholics and between the British and Irish states, as well as the changes in identity that have been analyzed in this research. What consociationalists do not recognize is that the transformation of conflict at the microlevel is inextricably linked with the reform of political institutions. Microlevel transformations and the reform of political institutions are not mutually exclusive options; rather, they may and indeed *should* take place at the same time (Ruane and Todd 2004, 1996; Dixon 2005; Bloomfield 1996).

Refusing to admit the possibility of microlevel transformations means that consociationalists also have overlooked the importance of the British government's civil society approach to conflict resolution. The result of these policies has been the creation of an officially sanctioned public sphere in which cross-community groups and projects are funded and supported by the government. Civil society groups that do not fit the government criteria face a comprehensive network of institutions that are meant to enforce the officially sanctioned public sphere, such as the Community Relations Council. Although the effectiveness of these institutions in carrying out their appointed tasks may be questioned, this research suggests that these structural changes are more important than the individual success stories of particular organizations or programs. Like it or not, these are new structures that civil society actors must work within. Civil society groups that "buy into" the logic behind the approach receive funding and government support; those that do not are forced to rebel against the system or adapt to it.

The evangelical response to this structural and institutional change indicates that there are no longer any "Queen's rebels" amongst them (Miller 1978). Mediating and postevangelicals have embraced the logic behind the civil society approach, as it dovetails with their own aims. Traditional evangelicals have adapted selectively to it, even though it does not correspond with their ideals. The extent that traditional evangelicals have adapted has not yet been recognized. The significance of their adaptation is that it signals a move away from what Mitchel (2003) would call "religious nationalism" to a focus on the "moral issues" that are such a prominent feature of evangelical politics in other contexts, such as the USA.

Indeed, a comparative analysis of evangelicalism in the USA, Canada, and Northern Ireland provides further insight into the linkages between evangelical identity, activism, and political institutions. For instance, American evangelicals employ interest-group politics and target the Republican Party, tactics that fit well with the USA's diffuse, federal political institutions. To the extent that a devolved Assembly and plans for more streamlined, accountable local government develop, political structures that favor interest-group and party-centered tactics could emerge in Northern Ireland. Traditional evangelicals, with their focus on moral issues and their foothold in the DUP, could thrive in such a situation. However, these strategies would not work well within a highly centralized political system, as in Canada. Canadian evangelicals have adapted to those political institutions by cultivating discrete relationships with politicians across party lines and with other civil society actors. This is the sort of relationship that mediating evangelicals developed with government ministers during direct rule. Those relationships depended, to some extent, on mediating evangelicals' enthusiastic acceptance of government's pluralism-promoting, civil society approach. To the extent that these sorts of relationships continue, there is a strong structural incentive for Northern Irish evangelicals to adapt along Canadian lines. Mediating evangelicals' sociopolitical projects have been facilitated by direct rule structures; with devolution, traditional evangelicals could come back into their own—especially with Paisley in the position of first minister in the Assembly.

Religion and Conflict Transformation: Policy Implications

Conflicts with a religious dimension require conflict resolution policies that recognize and address the religious dimension. Northern Ireland is no exception. This research has explored why addressing the evangelical dimension of the Northern Ireland conflict is a vital part of the overall process of transformation. Accordingly, it is possible to draw out specific policy implications for maximizing evangelicalism's transformative potential in Northern Ireland. This also contributes to the general theoretical and practical understanding of the role of religion in conflict transformation. Using the case of Northern Irish evangelicalism, we can identify broad principles that can help to maximize religion's transformative potential in other conflict situations. A comprehensive understanding of these processes and principles, and their policy implications, includes the following dimensions:

An understanding of the historical role of religion in various conflicts: Religious contributions to conflict resolution do not come in "one size fits all" packages. The significance of religious contributions to particular conflicts depends on the role that religion has played in sustaining the conflict. In Northern Ireland, evangelicalism contributed to boundary formation between Catholics and Protestants and underwrote unionist power. This historical legacy means that conflict transformation depends in part on the ability of evangelicals to break down boundaries and to reconceptualize their relationship with sociopolitical power. This research has demonstrated how mediating and postevangelicals have begun this process; it has further argued that the significance of their work should not be underestimated. These evangelicals should be encouraged to participate in the public sphere *as evangelicals*. This recognizes the integrity of their position and legitimates ideals that are widely respected in the broader Protestant community, such as esteem for the Bible. Their work can also provide evangelicals, and the wider Protestant community, with religious and moral justifications for embracing social and political change. In other conflicts with religious dimensions, a similar process could be encouraged. This would involve identifying how religion is important to the conflict. It would then require supporting religious groups that are actively reconceptualizing religion's relationship with conflict and devising peacebuilding alternatives.

An understanding of religion's relationship with power: Religious actors have more freedom to contribute to change when they do not have a close relationship with power. In Northern Ireland, evangelicalism's relationship with power has been breaking down. This has created the space for evangelicalism to play a "prophetic" role. This research has demonstrated that Northern Irish evangelicals have responded in diverse ways. On the one hand, mediating and postevangelicals have instituted extensive peacebuilding or antisectarian programs. On the other hand, traditional evangelicals have accepted the pluralist political order and have increased their focus on "moral issues." Both of these responses have the potential to contribute to the building of a postconflict society where religious actors—even if they disagree with each other or with other civil society groups—have their say in a "normal" democratic public sphere. In other conflicts with religious dimensions, religion's independence from sociopolitical power also should be encouraged.

An understanding of the structure of civil society: Government has a role to play in managing the potential for conflict amongst civil society groups. The extent that civil society is structured by government policies that promote pluralism, equality, and cross-community dialogue and initiatives is important. Religious actors, like other civil society groups, must negotiate those structures. In Northern Ireland, the British government's civil society approach has not solved every problem, but it has contributed to more effective conflict management. Mediating and postevangelicals have embraced these British government reforms and participated enthusiastically in the new structures, thus lending support and legitimacy to them. However, the new officially sanctioned public sphere still excludes some groups and risks alienating them. Some traditional evangelicals claim to be amongst the excluded, as they believe that their efforts to participate in the new structures have been systematically rebuffed. An effort should be made to include traditional evangelicals more fully in the process. Some of their criticisms of the approach and other specific policies have merit and reflect the views not just of traditional evangelicals but also of the wider Protestant community. In other conflicts with religious dimensions, a similar effort should be made to include the viewpoints of religious actors that both agree *and* disagree with ongoing reforms.

An understanding that theologies matter: It is not enough just to change religion's relationship with sociopolitical power. Religious contributions to conflict resolution must also have a viable theological basis that appeals to the deeply felt religious sensibilities of the people. In some cases, this may mean that aspects of religious belief that have been used to justify conflict or to reinforce oppositional identities must change. This process is most effective if indigenous religious actors critique aspects of their theology that have contributed to conflict and articulate theological alternatives that focus on concepts such as forgiveness, reconciliation, or peacebuilding. In Northern Ireland, mediating and postevangelicals have begun to do this through educational programs for both clergy and laypeople. Secular peacebuilders who dismiss the importance of theological justifications for change or peacebuilding do a disservice to potential allies. If they aggressively attempt to exclude religion from the public sphere, they undercut the work of religious peacebuilders and risk alienating potentially large swathes of the population. In conflicts with religious dimensions, excluding religious language from the public sphere may unnecessarily eliminate the construction of a potentially transformative new vocabulary. A *reemphasis* on religious aspects of difference—but from a pluralist rather than ethnonationalist perspective—could contribute to the transformation of conflict.

Evangelicals are seeking but have not yet found their place in post–Belfast Agreement Northern Ireland. As evangelicals have negotiated the new dispensation, new evangelical identities and new evangelical sociopolitical projects have emerged. These new identities and projects are a product both of individual agency and reflection and of changing sociostructural conditions. The future development of evangelical politics depends on the theologies and strategies promoted by evangelical activists. It also depends on the willingness of government and other civil society groups to include traditional and mediating/postevangelicals in the public sphere. Evangelicals from all points on the spectrum have the potential to contribute to a post-conflict society, but this potential will not be realized if their efforts are overlooked or their concerns ignored.

Appendix

This research was designed to analyze evangelicals' ongoing and potential contribution to the transformation of conflict in Northern Ireland. This appendix provides an overview and justification of the ethnographic methods. It presents a detailed account of the research process, including explanations of how interview questions were devised and how the interview data was coded and analyzed. It also addresses problems such as how a researcher gains access to a relatively "closed" community such as Northern Irish evangelicals, and how the researcher's identity impacts the process of gathering and analyzing the data.

Research Methods

The selection of my research methods was driven by my concern to take into account the relationship between structures and religious beliefs. Social and political structures provide the context within which people act, but they cannot tell us exactly how they will act or why they will act that way. In order to fully understand processes of change, we need to include a "bottom up" approach that focuses on the stories that people tell about how they perceive themselves, experience change, justify their social activism, and so on (Ammerman, 1994; Neitz, 2004). The best way to access these stories or narratives is through an ethnographic approach that allows people the maximum freedom to express themselves.[1] Accordingly, I used methods such as semistructured interviews and participant observation.[2] These methods are well-suited for gathering data that answer questions about *how* microlevel processes take place. Open-ended interview questions encourage participants to construct detailed narratives. This gives them the scope to talk at length about what seems most important to them.

It is important to be clear about the kind of data generated by these methods, and why it was important to gather this kind of data.

Interviews are time-consuming and in-depth, which necessitates a small sample size. I conducted 61 interviews with 57 evangelicals. They lasted from one to three hours. This means the data is not easily generalizable to the larger evangelical population, as may be the case in large-scale surveys. But it provides three sorts of insights that cannot be gained through quantitative research methods.

First, the open-ended nature of semistructured interviews means that the researcher avoids imposing rigid frameworks on the participants, as may occur in large-scale surveys or quantitative interviews. This means that researchers may often be surprised at their findings, as participants offer them interpretations that they had never thought of before. This happened on several occasions in my interviews.[3] Second, semistructured interviews capture participants' perceptions and thought processes in all their complexity and contradictions, which allows for movement beyond oversimple typologies and explanations. This sort of data cannot be generated in large-scale, quantitative research. In addition, participant observations recorded in field notes provide valuable background information, confirm participants' interview responses, or provide the researcher with new hypotheses to explore. Participant observation is also vital for building the sort of relationships that make the interview process both open and enjoyable. Third, the data generated by these methods can answer the all important "how" questions. Quantitative methods also answer how questions, of course, but data generated by qualitative methods have the advantage of keeping the focus on *participants'* perceptions of how processes occur. This keeps the researcher from "explaining away" participants' perceptions—something that has occurred all too often in research on religion (Spickard, Landres, and McGuire 2002). It also allows the researcher to discover competing perceptions of reality, both amongst different participants and within the same participant. Finally, the insights gained from understanding how processes occur may alert the researcher to changes that are occurring at the microlevel and have the potential to become more widespread. Quantitative methods may not pick up these nuances.

Selecting the Sample

Once I had identified the appropriate methods, I had to select a sample of evangelicals who were willing to participate in the research. This process was theory-driven. Drawing on Marty's (2000) work on religious structures, I decided to center my research in congregations and special-interest organizations. Congregations are likely to be places where socialization (including identity change) occurs, while special-interest organizations are the prime vehicles for nongovernmental politics (including the reframing of sociopolitical projects).

Once I had gained access to the field (see "Gaining Access," below), I relied on "gatekeepers" to direct me to participants. They helped me to select a sample that represented the range of views within the broad categories of traditional and mediating evangelicalism (see chapter 1). The selection of each congregation was based on its representative position on the traditional-mediating spectrum and its geographical location (urban or rural). The organizations also were selected on the basis of their position on the traditional-mediating spectrum. I was able to ascertain their positions through my own observations and my relationships with gatekeepers. Selecting congregations from urban and rural locations allowed for analysis of how the environment in which a congregation is situated impacts upon it. I included Zero28 and ikon in the congregational sample and analysis because these organizations had some distinctly congregational characteristics. Their narratives also dealt directly with many of the same issues of identity and identity change that were present in the narratives of people from the congregations.

Within the congregations, the pastor served as the gatekeeper. I asked each pastor to put me in touch with five to seven people from his congregation of varying ages, education levels, occupations, gender, social class, and theological and political beliefs. I asked that the sample include at least some people in leadership positions. This was meant to capture a *range* of views within the congregation. Although this runs the risk of the pastor-gatekeeper providing me with participants who agreed with him or who would present an uncritical picture of the congregation, or of him filtering out people with views he considered "unrepresentative" or "undesirable," I do not think that this was the case. Rather, the pastors directed me to a range of individuals whose views differed considerably from one another and from their own. For instance, after I had conducted my initial interviews in the rural Presbyterian congregation, I realized I had only talked to one person with a traditional identity. Since the pastor had a traditional identity and I suspected that there were more people in his congregation with traditional identities, I asked the pastor to direct me to two more people with views that were closer to his own. The participants from the urban Presbyterian congregation were all critical of what they perceived as the congregation's lack of focus on peacebuilding issues. And in the urban Free Presbyterian congregation, my initial sample included only one woman, so I asked the pastor to direct me to another. The participants spoke freely about the range of views within their congregation and other churches in the area. There may have been other views represented in each of these congregations; however, the participants did not mention them. Nor was there other internal or external evidence that I was directed to a biased sample. I was, therefore, confident that I would achieve my objective to investigate the range of views and the ways in which people constructed different positions. The data is not necessarily representative of evangelicalism, but that was not the aim of the research.

This process yielded an overall congregational sample of 41 participants (36 church members and five pastors).[4] Of the 36, there were 14 mediating evangelicals, 9 traditional evangelicals, 7 postevangelicals, 5 pietist evangelicals, and 1 who did not identify herself as a Christian. There were 23 males and 13 females. This imbalance reflects the predominance of males in leadership positions within the congregations. As far as age, thirteen were under 35, nine were between 36 and 50, nine were between 50 and 65, and five were over 65. The greater proportion of those under 35 reflects the youthful membership of Zero28 and ikon. In class categories, 24 were from the middle class, eight were from the working class (including blue and white collar workers), five were pastors, three were students, and there was one homemaker. The middle-class category included a range of occupations, from shopkeepers to community workers to higher professionals. The middle classes were overrepresented because of Zero28 and ikon (which was almost exclusively middle class) and the urban Presbyterian church, which was located in an upper-middle-class area of the city. Of the five pastors, two had traditional identities and three had mediating identities. All of the pastors were male. Taken together, the congregational sample represents the broad spectrum of evangelical identities within Northern Ireland.

The selection of the organizations was based on their broadly representative positions on the traditional-mediating spectrum. I gave priority to organizations that were well established or relatively well known within the public sphere. I was able to ascertain each organization's broad position through a careful reading of their literature and Web sites, my own observations, and my relationships with gatekeepers.

Within each organization, I sought to talk with staff members and activists (such as people who participated on steering committees). For the traditional organizations, I relied on a gatekeeper who had been introduced to me by a pastor in my home state of Maine. He directed me to five other participants who were active in three traditional organizations. I conducted a total of eight interviews with traditional activists (two were follow-up interviews with the same activists). For the mediating/postevangelical organizations, I relied on a gatekeeper I met at the Evangelical Contribution on Northern Ireland (ECONI) Sunday School and other contacts I had met in evangelical circles in Northern Ireland and the Republic of Ireland. I conducted a total of 22 interviews with mediating/postevangelical activists in four organizations (one was a follow-up interview with the same activist). I talked to more mediating/postevangelical than traditional activists because their organizations were larger and had less overlap in staff and membership. Taken together, this sample represents the broad spectrum of Northern Ireland's evangelical organizations.

The total of 61 interviews allowed me to reach a "saturation point" in which the spectrum of Northern Irish evangelicalism was adequately explored. My confidence in the sample was bolstered by my participant observation, my ongoing review of the literature on Northern Irish evangelicalism, and the data gathered during my pilot study from 2000 to 2001 (this included 20 interviews; see Ganiel 2002).

The interview questions for participants in the congregations and the organizations were shaped by the idea that the primary functions of congregations and organizations are different. The questions I devised took this into account, although there was some overlap. The open-ended nature of the questions meant that the data that was produced was well suited for qualitative narrative analysis. Qualitative analysis of narratives involves a systematic process of coding or filing chunks of texts round themes or codes that are generated from the interviews. It differs from quantitative methods of narrative analysis (such as some forms of content analysis), which code data by, for example, counting the number of times specific words appear in texts. Qualitative narrative analysis is a more effective means of preserving the contextual nature of the stories people tell, without judging their significance on crude word counts (see Lofland and Lofland 1984; Schatzman and Strauss 1973).

The following traces the steps I took in designing the interview schedule and in analyzing the interview transcripts. I also took comprehensive field notes, to which I referred for background informationnand in which I recorded confirmatory and surprising observations. Documenting and justifying this process establishes the validity, credibility, and reliability of the data and the analysis.

Congregations, Identity, and Change

The questions for people in the congregations were designed to generate data that was broadly centered around the process of socialization, including identity formation, construction, and change. For example, I sought to establish the content of people's evangelical identity, and what factors had contributed to the formation of that identity.

I established the way participants constructed their identities by asking open-ended questions about their conversion experience, what it meant to be a Christian citizen, if the church or clergy should be involved in politics, what the relationship between Christianity and politics should be, cross-community activities, their thoughts on Catholicism, the possibility of a united Ireland, the Belfast Agreement, the Orange Order, and whether or not they supported a particular political party.[5] I did not ask specifically about the use of violence, but many participants condemned the violence of paramilitaries on both sides, unprompted. I also asked questions about the relationship between church and state, and pluralism (including questions about cross-community activities and ecumenism). These questions produced narratives in which participants addressed key themes of their systems of belief and action. This allowed me to construct the empirical types (see chapter 1). Their narratives were multilayered and complex, and participants slipped between religious and political themes as they spoke, sometimes developing those themes in unexpected ways. I coded these responses under the broad theme of "identity."

I also was keen to discover if there had been changes in participants' identities, or in their religious or political beliefs. I asked them directly about change; if they said they had experienced change, I asked them to identify the factors or experiences that had contributed to it. These questions produced narratives that described dramatic experiences or epiphanies in which the participants had radically changed course. They also produced narratives describing slow, evolutionary change. Some people claimed that they had not changed at all. I coded these responses under the broad theme of "perceived change."

Finally, I wanted to understand how participants' identities or changes in their identities impacted the way they thought about social or political activism. Questions that often produced answers that dealt with this theme were about the role of the church in society and politics, Christian citizenship, the role of clergy in politics, their own participation in their congregation or in special-interest organizations, and the challenges facing the church. I coded those narratives under the broad theme of "meaning given to sociopolitical activism." After coding the data, I grouped the texts together under the various codes and compared the narratives both within and between congregations.

Organizations and Nongovernmental Politics

The questions for people in the special-interest organizations were designed to generate data that was broadly centered around the practice of nongovernmental politics. I wanted to establish what sort of activities the organizations were involved in, including their strategies and their goals. I also wanted to find out how activists framed or gave meaning to their participation in the organizations, including how they justified their organizations' activities.

To that end, I asked direct questions about what the organizations do. These questions varied from activist to activist, especially if the participants played a different role within the organization (staff member or volunteer, for example). These questions generated responses that provided comprehensive lists of the organizations' activities. The organizations' Web sites and literature provided further information of this nature. These questions also generated responses that explored how the activists gave meaning to their participation in the organizations. These answers included personal stories, where the activists linked their life experiences and personal changes with their activism. I asked additional open-ended questions that invited participants to evaluate their organization, the role of the churches and evangelical organizations in civil society, their relationships with other organizations, and the "new" public sphere ushered in by the Belfast Agreement, amongst other things. These sort of questions generated responses dealing with how activists justified their reactions to change.

I went about the analysis by coding the data around the two themes of "what the organizations do" and "how activists give meaning to their activism." The "what the organizations do" category was broken down further into the following subcodes: founding, activities and operation, networks, funding, change, evaluation, challenges, and government interaction. The "how activists give meaning to their activism" category was broken down into the following subcodes: relationship with Catholicism, relationship with evangelicalism, relationship with the churches, church in society and politics, the Northern Ireland context, the wider context, cross-community, marginalization, and moralism. I grouped the texts together under the various codes and compared the narratives both within and between organizations.

The research amongst the congregations and special-interest organizations was set within the broader context of religious networks. In the case of the special-interest organizations, I deliberately selected organizations that I knew to be "networked" with each other. I also asked people in the special-interest organizations if they cooperated with other organizations or congregations as they pursued their goals. I did not deliberately seek out congregations that I knew to be involved with the organizations in the traditional or mediating networks. But I did ask people in congregations whether or not they or their church participated in any evangelical organizations.

The coding process was not as straightforward and tidy as this account may make it appear. For instance, some texts explored multiple themes and thus were coded under multiple categories. The association of socialization with congregations and of nongovernmental politics with organizations was not always clear-cut. In addition, the postevangelicals from ikon and Zero28 did not fit comfortably within either the congregation or special-interest organization category. They defined themselves as a "community" and resisted calling ikon and Zero28 either a "church" (congregation) or an organization. My interviews with them were much less structured, although they still addressed the broad themes of socialization and nongovernmental politics. Ikon and Zero28 are included in both the chapter on congregations and the chapter on organizations.

Gaining Access

Seeking out a sample that is likely to include at least some of each empirical category is not a simple task. I had to be not only familiar with the nuances of Northern Irish evangelicalism (for example, to identify the difference between traditional and mediating congregations and organizations) but also capable of gaining access to evangelicals. My identity as both an "insider" (an evangelical) and an "outsider" (an American) aided me in my pursuit.

With the "cultural" or "reflexive" turn in ethnography, the self has been conceived of as relational.[6] This means that people are conceived of as constructing their identities and presenting themselves in relation to others. Much of the debate in contemporary ethnography focuses on how a researcher's identity as an insider or an outsider affects the participants' presentation of themselves. The insider/outsider distinction is especially complicated in the study of religion (Ganiel and Mitchell 2006; Spickard, Landres, and McGuire 2002; Arweck and Stringer 2002; Nason-Clark and Neitz 2001), and in the study of religious groups that could be considered closed, such as evangelicals. Here, access is an important consideration. Evangelicals are the stuff of stereotypes, with reputations as Bible-thumping, closed-minded hillbillies. They are keenly aware of their negative public perception and might be wary of "outsiders" who come to study them. They also inhabit a clearly defined subculture (Smith 1998; Warner 1988) that might be difficult for outsiders to negotiate. Insiders, on the other hand, might find it easier to gain access to their coreligionists. But their conclusions might be viewed with suspicion by social scientists, especially if they claim to have an exclusive understanding of the group or engage in advocacy.

If penetrating the evangelical subculture has been deemed an arduous task, nowhere is this more evident than in Northern Ireland. It seems that the religious dimension of the conflict has made evangelicals even more wary of researchers. Bruce (1986), an outsider as a secular social scientist, complained that it was hard to get information out of ordinary Free Presbyterians. Cooke, a Methodist minister, was refused point-blank when he asked to interview members of Ian Paisley's Free Presbyterian Church (1996:115). Ronson (2002), a secular Jewish journalist, has provided a high-spirited (but cautionary) account of following Free Presbyterian ministers Ian Paisley and David McIlveen on a missionary trip around Africa. Ronson was unable to get Paisley and McIlveen to talk about politics and was keenly aware that he was viewed with suspicion.[7]

With Bruce, Cooke, and Ronson's tales in mind, I approached the field with some trepidation. However, I found that gaining access was surprisingly painless. I believe this was due largely to my background in American evangelicalism and to the people who put me in touch with Northern Irish evangelicals. The "gatekeepers" who helped me to find participants for the interviews were respected within evangelicalism. Participants trusted them and were in turn willing to talk with me.

I took two routes into Northern Irish evangelicalism: one through the Republic of Ireland and the other through the USA. At the time of the research, I lived in Dublin and attended a Baptist church. As I shared information about my research with my pastors and others in the congregation, they gave me literature about Northern Irish evangelicalism and offered to put me in touch with key people, such as the staff of ECONI and the International Fellowship of Evangelical Students in Ireland. These contacts enabled me to conduct a pilot study of Northern Irish evangelicalism from 2000 to 2001 (Ganiel 2002). The people I met through that study and at various evangelical events I attended helped me when I began this research project in 2002. For instance, at the 2001 ECONI summer school I met the rector of the Church of Ireland parish that participated in this study, as well as an ECONI activist who facilitated my research on ECONI and on the urban Presbyterian congregation that participated in this study. These gatekeepers provided me with access to mediating evangelicals.

My other route was through my membership in an Orthodox Presbyterian Church (OPC) in Bangor, Maine. The OPC is a small, conservative denomination that broke away from the mainline Presbyterian Church-USA in 1936. Another pastor in the denomination is from Northern Ireland, and he put me in touch with an activist who gave me a great deal of help in contacting other evangelicals. This gatekeeper provided me with access both to traditional congregations and organizations.[8]

When I approached people about participating in the research, I did not always identify myself as an evangelical. Sometimes I volunteered the information, sometimes people asked me about my church background, and sometimes it was not mentioned at all. I do not think anyone used my evangelical identity as a test as to whether they would talk to me or not. Most seemed assured by the channels through which I had approached them and were happy to help out an American student. However, if I had not myself been attending evangelical churches, I would not have been able to approach them through the same evangelical channels. That would have made access much more difficult.

Relationship with Participants

The data-gathering process depends in large part on the relationships between the researcher and the participants. Before the cultural or reflexive turn, the accepted wisdom was that researchers should remain aloof from their subjects so that they could offer "objective" analyses. Now, however, this paradigm has been replaced with an understanding of how the researchers themselves affect the entire process of the production of information (Clifford and Marcus 1986; James, Hockey, and Dawson 1997; Brewer 2000). Researcher and participants work together to produce the data, and how they perceive each other impacts the data that is generated.

The researcher's identity is important during this process. For instance, participants might be more comfortable and open with a religious insider. Insiders may share a common set of assumptions and a distinct religious vocabulary. Outsiders, on the other hand, may have difficulty getting past those assumptions and that vocabulary (Warner 1988). They may be seen as candidates for conversion attempts (Gordon 1987; Peshkin 1984). As a religious insider, I expected that the participants would freely share their thoughts with me. As an ethnic or national outsider, I was unsure how the participants would react. In this case, I found that they were always keen to explain themselves carefully, often going into great detail to make sure that I understood things that they thought an American would not understand. Many drew parallels between Northern Ireland and the USA to illustrate their stories.

However, it is simplistic to assume that clear binaries between religious insider and outsider always actually exist, or that research participants conceive relationships in these terms (Ganiel and Mitchell 2006; Neitz 2002; M. Wilcox 2002). This can be illustrated by considering the differences between the liberal and conservative branches of evangelicalism.[9] If a researcher presents himself or herself as a religious insider at the beginning of a research relationship, information the researcher reveals during the course of the interaction may change the participant's initial perceptions. The researcher may expose himself or herself as too liberal or too conservative. For example, if I told a Northern Irish evangelical that I was a member of the OPC, they might make assumptions about my beliefs. In the cases in which I mentioned the OPC, many Northern Irish evangelicals knew about the controversies that led to its formation, and the writings of its founder, J. Gresham Machen.[10] Some identified it with the Evangelical Presbyterian Church (which broke away from the Presbyterian Church in Ireland at around the same time); one even said that its teachings were not "too far" from the Free Presbyterian Church. Such identifications may have made conservative Northern Irish evangelicals more comfortable with me, and liberal Northern Irish evangelicals uneasy. My Baptist identity, on the other hand, was quite ambiguous because of the broad liberal-conservative spectrum amongst Baptists.

I expected the insider/outsider and conservative/liberal evangelical boundaries to be more important during the research process than was actually the case. I expected most evangelicals to ask me early on if I was "saved" or "born again." This did not happen very often. Sometimes my salvation status was established during the interview, after it, or not at all. This led me to believe that most of the evangelicals would have told me exactly what they did, whether or not they thought I was a religious insider.[11]

The conservative/liberal distinction was important in a few cases. For instance, one participant did not immediately accept my presentation of myself as a believer. He asked a number of follow up questions about my beliefs and where I went to church in Dublin and the USA. The OPC seemed acceptable because it had broken away from the more liberal Presbyterian Church USA. He was also pleased that my congregation in Maine used the Reformed Trinity Hymnal; he said that because I was saved, I would be better able to understand his answers to my questions.

On the other hand, those who could be placed on the liberal end of the spectrum of evangelicalism never asked me questions that seemed designed to determine if I was *really* saved. With those evangelicals, my experience was more like that of Guest (2002). In his research with postevangelicals and alternative worship communities, Guest found that his participants did not label him in any particular way. I felt quite free to converse with these evangelicals, both during formal interviews and socially.

At times it seemed like my insider identity was more problematic for *me* than for my interviewees. For instance, when talking with conservative evangelicals, I was careful not to say anything that would indicate that I disagreed with some of their positions. This may have led them to believe that I shared more of their beliefs than I actually did, especially if they made assumptions about my beliefs based on their knowledge of the OPC. I felt guilty because I might be misleading the participants. To offer an authentic self-presentation, I tried not to identify too closely with certain religious beliefs or dress codes. For example, in my research amongst Free Presbyterians, I did not observe conventional Free Presbyterian dress codes. Free Presbyterian women wear dresses and hats to church. I did not wear hats to Free Presbyterian services, and on one occasion I received communion while wearing trousers.[12] I became confident that I was accepted as an insider when, on several occasions, even conservative evangelicals offered to help me find a husband. It is generally taboo for evangelicals to marry outside of evangelicalism.

However, I came to believe that my identity as a religious insider had no more impact on the data-gathering process than some of my other identities. My national identity often seemed the most important. Being an outsider in this sense generally eased the research process, as participants assumed I had little knowledge of the local situation and were keen to make themselves clear. They sometimes seemed grateful for the chance to explain themselves to the outside world. Interestingly, many assumed I was Canadian. Initially, this was surprising, but as more people made the mistake, I attributed it to several factors. Having grown up less than an hour from the Canadian border in Maine, it is possible that I have a slight Canadian accent. Or, given that Ulster Protestants immigrated in proportionally larger numbers to Canada than to the USA, they may have assumed that I was returning to Ulster to study my ancestors. Northern Irish Protestants, understandably, have been known to assume that Irish-Americans are Catholic supporters of the Irish Republican Army. I was glad not to be associated with this Irish-American stereotype and felt it was a good thing to be perceived initially as Canadian. Once my national identity was established, participants often drew parallels between events in Northern Ireland and events in the USA. Many spoke fondly of America's Puritan forefathers, and the many American presidents of Ulster-Scots descent. Being young, female, and a student also seemed to put the participants at ease.

All these factors combined to make the data-gathering process relaxed and enjoyable. I was invited into homes, given lifts all over the countryside, and fed so many scones and biscuits that an expanding waistline was a far more pressing worry than convincing evangelicals to talk to me. As such, the data that was generated was rich, detailed, and reliable.

Researcher Identity and Data Analysis

The researcher's identity is important when it comes to analyzing the data. Here, the insider-outsider debate comes back into play. For instance, a researcher who is a religious insider may be specially equipped to be able to understand and interpret religious groups. Stringer (2002) has put forward this interpretation, using Cantwell Smith's notion that there is something distinct about religion that is different from other identities: faith. A shared faith does not have to be drawn from one particular religion but refers to the experience of faith in general. Stringer concludes that faith is an essential concept and cannot be fully understood by the nonreligious. This leads him to argue that insider studies of religion are to be encouraged, although they can also be effectively conducted by "empathetic" outsiders if they acknowledge and respect the need for mystery. In addition, secular researchers may be more likely to "explain away" supernatural explanations or stories offered by participants. Although supernatural experiences cannot be objectively verified, participants' perceptions of these experiences impact their actions. As such, discounting participants' self-perceptions may blind the secular researcher to the dynamics of some sociopolitical processes.

On the other hand, it may be tempting for the insider to accept the participants' perceptions and conclusions at face value, or to slip into advocacy. They may place too much importance on theological or spiritual interpretations or explanations and discount sociological or political explanations. In a recent study of Northern Irish evangelicalism, evangelical Patrick Mitchel[13] seems to slip into advocacy in the conclusion of his book, implying that ECONI may be equipped to bring about broad, positive changes in Northern Irish society:

> It remains to be seen how effective a relatively marginal organization [ECONI] will be in catalysing a broad re-evaluation of attitudes within Ulster evangelicalism during the twenty-first century. It is ironic, but quite consistent with the 'foolishness of God' who demonstrates his power through apparent weakness (I Cor. 1:18–2:5), that the waning political power of Ulster evangelicalism may prove in time to be a spiritual blessing. (2003:317)

As an evangelical insider, Mitchel writes not only for a general audience that seeks to understand Northern Irish evangelicalism, but also for *evangelicals*. He wants to contribute to their understanding of what the church is and what it *should* be doing in Northern Ireland. The large volume of literature produced by ECONI over the last 15 years is written in a similar vein.[14] Evangelical historian Mark Noll also writes with this dual audience in mind (2001a, 2001b, 1995). Both Mitchel and Noll take care to separate their scholarly conclusions from their normative concerns. Religious insiders too must always be careful to do so.

As an evangelical "insider," I tried to separate my scholarly conclusions from my normative concerns. However, I am uncomfortable with the forced binary behind the concept of a dual audience. While I believe it is possible to separate myself from my evangelical background for the purpose of sociological analysis, I wonder if there are not some underlying assumptions at work. Noll's and Bruce's differing interpretations of secularization help to illustrate my concern. Both Noll and Bruce recognize that secularization is taking place in the sense of decreasing church membership and attendance in Western nations, and so on. But Noll (2001a, 2001b) argues that this could lead to a renewal for Christianity, while Bruce concludes that it will lead to the eventual eradication of Christianity (1995, 2001a). Although reluctant to speculate on Noll's and Bruce's underlying assumptions, I wonder how Noll's faith and Bruce's lack thereof influenced these interpretations. My doubts, however, have led me to take particular care when analyzing my data. Accordingly, I have been open about my evangelical background, have made sure that my accounts of how the research was carried out are transparent, and have focused on research questions that ask *how* sociological processes take place.

Notes

Chapter 1   Introduction

1. The official name of the agreement is the "Agreement Reached in the Multi-Party Negotiations." In public discourse, it is usually referred to as either the Belfast Agreement or the Good Friday Agreement.
2. For analyses of the DUP's emotive discourses and their possible implications for the peace process, see Ganiel (2007) and Rankin and Ganiel (2007).
3. For analyses of how religion has mattered in Northern Ireland, particularly for Protestants, see Bruce (1986), Akenson (1992), Fulton (1991), Buckland (1981), Heskin (1980), O'Malley (1983), O'Brien (1974), Rose (1971), Todd (1987), and Wright (1973).
4. See also the five-volume Fundamentalism Project by Marty and Appleby (1992–1995).
5. For a detailed account of the methods, see the appendix.
6. This does not necessarily mean that the "real people" concerned use the terms "traditional" or "mediating" to identify themselves. However, this does not preclude them from identifying with these terms, either.

Chapter 2   Civil Society, Religion, and Conflict in Northern Ireland

1. Marx viewed civil society as an arena for oppression, in which the dominant classes controlled the masses.
2. Keane defines postfoundationalism as the rejection of the search for a "good society" based on universal (liberal) ideals. This is a position similar to that taken by radical democrats.
3. Some of the literature emphasizes the durability and lack of change in identities in Northern Ireland; According to Hayes and McAllister, (1999:30–48) this is because they work with macro categories of identity. See also McGarry and O'Leary (2004). For a critique, see Ruane and Todd (2004).

4. Kymlicka writes out of the Canadian context, a "multinational" society that includes Quebec. Parekh writes out of the increasingly multicultural UK context, with a definition of culture that carries assumptions about the validity of British social and political institutions.
5. It is not inevitable that state or established religions will always play a priestly role. See Davie (1994) on how the Church of England played a prophetic role with its "Faith in the City" report, which challenged the Conservative government's economic policies.
6. According to the Northern Ireland Voluntary and Community Almanac: State of the Sector III (2002), compiled by the Northern Ireland Council for Voluntary Action (NICVA), there are 4,500–5,000 voluntary organizations in Northern Ireland. NICVA is located in the Department of Social Development (DSD) and has about 1,000 affiliated members.
7. With the intensification of the Troubles came a corresponding growth of grassroots PCROs. In 1968, there were only a handful of "organisations within a fairly traditional peace movement," but by 1993 there were more than 70 PCROs listed in the CRC's *A Guide to Peace and Reconciliation Groups* (Wilson and Tyrrell 1995:245). According to the CRC Web site, there are currently more than 130 community relations groups (http://www.community-relations. org.uk, accessed April 16, 2005).
8. The Commission (chaired by Norwegian human rights lawyer Torkel Opsahl) was an independent, international commission that invited citizen discussion about the future of Northern Ireland. See Pollak (1993). Farrington (2004b) argues that the Opsahl Commission was instrumental in creating a public sphere in Northern Ireland.
9. Such as members of the Women's Coalition or former loyalist paramilitaries such as Billy Hutchinson and David Ervine of the PUP.
10. The British government introduced the Fair Employment Acts in 1976 and 1989, which included a Fair Employment Agency. After the Belfast Agreement, an Equality Commission was introduced (http://www.equalityni.org). The Equality Commission replaced the Fair Employment Commission, the Equal Opportunities Commission, the Commission for Racial Equality, and the Northern Ireland Disability Council. See McCrudden in Ruane and Todd (1999).
11. The British government's broader project has attempted to narrow the economic, social, and cultural gap between Catholics and Protestants through measures such as Fair Employment and public recognition/ "parity of esteem" for identities (Thompson 2002). These include provisions of the Belfast Agreement such as the Equality Commission and the Northern Ireland Human Rights Commission (NIHRC), as well as Section 75 of the Northern Ireland Act. There is evidence that fair employment and equality measures have contributed to the narrowing of the gap (McCrudden 1999; Osborne and Shuttleworth 2004). However, the NIHRC has come under severe criticism for

failing to meet its objectives, which included reviewing the effectiveness of laws and practices, making recommendations to government,
and promoting awareness of human rights. For instance, it was charged
with drawing up a bill of rights for Northern Ireland, but this project
collapsed due to internal divisions and lack of support from the British
government (Livingstone and Murray 2005). Section 75 of the
Northern Ireland Act required public authorities "to promote equality of opportunity" and "good relations" between people of different
religious belief, political opinion, racial group, age, marital status,
sexual orientation, gender, and disability.

12. The EU's Special Peace and Reconciliation Programmes poured
    more than £1.5 billion into peace and reconciliation work.
    Partnerships were required to include NGOs; local councilors; and
    representatives from the community, business, trade unions, and
    statutory sectors (See Taylor 2001:44–45 and Brewer 2003a:133).

13. The CRC will evaluate groups' and District Councils' community
    relations programs, and if they do not measure up, it will have the
    power to withhold funding. It is not clear how the CRC itself might
    be evaluated.

14. The written responses can be accessed at http://www.asharedfutureni.
    gov.uk.

15. A taskforce was first recommended in the 2000 "Consultation
    Document on Funding for the Voluntary and Community Sector,"
    or Harbison Report. See http://www.taskforcevcsni.gov.uk.

16. This was reflected in the largely cross-community character of orga-
    nizations that responded to "A Shared Future" (See responses to the
    document archived on http://www.asharedfutureni.gov.uk).

17. The Civic Forum should be seen also as part of a wider trend in the
    management of devolved government in the UK. For instance, the
    Scottish Executive and Parliament are formally supporting and funding
    a Civic Forum, a Civic Forum has been proposed for the new London
    Authority, and legislation for the Welsh Assembly requires a "formal
    consultation scheme with the voluntary sector" (I. Lindsay 2000:404).

18. This group is chaired by Lord Eames, the former Anglican Archbishop
    of Armagh, and Denis Bradley, who was vice chairman of the Policing
    Board. Other members include David Porter, the director of the Centre
    for Contemporary Christianity in Ireland (formerly Evangelical
    Contribution on Northern Ireland); James Mackey, a former philoso-
    phy lecturer; Elaine Moore, a drugs counselor; Rev. Lesley Carroll, a
    Presbyterian minister; Jarlath Burns, a former Gaelic Athletic
    Association player; and Willie John McBride, a former British and Irish
    Lions rugby player.

19. Some attempts to cultivate "public spheres of controversy" have orig-
    inated at the grass roots, including the Healing Through Remembering
    project and the "long march" from Londonderry to Portadown to
    publicize victims' rights. The sisters of Belfast man Robert McCartney,

who was murdered by members of the IRA in February 2005, could be said to have created a public sphere of controversy through their high-profile efforts to have his accused killers processed through the court system.

20. In 2005, ECONI changed its name to Centre for Contemporary Christianity in Ireland (CCCI). Since most of this research was carried out before the name change, the organization is referred to as ECONI throughout.

Chapter 3   Religion in Transition—
Comparative Perspectives

1. For a historical account of the Orange Order, see Haddick-Flynn (1999). For contemporary accounts, see Kaufmann (2007) and Dudley Edwards (1999).

2. Catholics also interpreted it as God's judgment, albeit a judgment that they were not Catholic enough (See Larkin 1976).

3. In the words of James Craig, prime minister of Northern Ireland from 1921–1940: "In the South they boasted of a Catholic State. All I boast of is that we are a Protestant Parliament and a Protestant State" (quoted in Dixon 2001b:50).

4. Of course, evangelicalism was not as unified as it may appear in this account. On occasion evangelicals and other Protestants "dissented" from the dominant Calvinist-inspired position (Ganiel 2003; McBride 1998).

5. Except, for example, a period of violence as partition bedded down (1921–1922) and the violence at the start of the Troubles in the late 1960s.

6. Republican paramilitaries were also opposed to Sunningdale and were not included in its negotiation.

7. See also McGarry and O'Leary (2004) and Tonge (2005) on the importance of including an "Irish dimension" to conflict resolution. Although original theories of consociationalism (as developed by Lijphart 1977) did not include "external" dimensions, these commentators see the external Irish dimension as necessary for the working of consociational institutions in Northern Ireland.

8. "Sunningdale for slow learners" has become a catchphrase for the Belfast Agreement; it is attributed to the SDLP's Seamus Mallon.

9. Paisley founded the Free Presbyterian Church in 1951, but it experienced its most rapid growth shortly after the beginning of the Troubles.

10. Although Northern Ireland is experiencing some secularization, it maintains higher levels of religious beliefs and practices than other Western societies. See Brewer (2002) and Fahey, Hayes, and Sinnott (2004).

11. As with Calvinism, there is no single Anabaptist tradition. Anabaptism had its origins in Reformation era Europe (especially Germany), when its insistence on the separation of church and state challenged both Catholic and Reformed models of church-state relationships (at that time, the Catholic and various Reformed churches advocated established churches). Anabaptists also endorsed pacifism and practiced adult baptism by immersion, rather than by infant baptism. These deviant beliefs resulted in their persecution by Catholic and Reformed churches, and many Anabaptist groups retreated from society to form their own communities. The Amish in the USA and various Mennonite communities around the world are the remnants of this radical form of Anabaptism. Contemporary Baptist denominations, which continue to advocate adult baptism and the separation of church and state, are also heirs of the tradition. Roger Williams, who was driven out of Puritan Massachusetts because of his Anabaptist tendencies, founded present-day Providence in Rhode Island as a colony that officially separated church and state, a model that would later be adopted in the U.S. Constitution. In 1920, Mennonites in North America developed the Mennonite Central Committee (MCC) as a vehicle for engaging more with the world around them, rather than withdrawing into isolated communities. The MCC has been active in sending peacemaking missionaries to conflict areas around the world, including Nicaragua, Somalia, South Africa, and Northern Ireland (ECONI often has had a Mennonite volunteer on staff) (See Appleby 2000:143–150).

12. Historically, some groups in the Anabaptist tradition, such as the Amish, have tended to withdraw from society, setting up alternative religious communities. Some contemporary Anabaptists have rethought this position and argued for a more focused engagement with the society (Appleby 2000). Yoder's *Politics of Jesus* (1994) might be considered a call to engagement, as well.

13. ECONI is not purely pacifist in the Anabaptist sense. Some evangelicals within it would adhere to some versions of just war theory.

14. Brewer's (2003b) analysis of grassroots peacemaking amongst Catholics and Protestants has a similar ecumenical focus. See also Brewer, Bishop, and Higgins (2001).

15. There have been strong strains of evangelical anti-Catholicism in the USA and Canada. For more on anti-Catholicism in North America, see Shea (2004), Massa (2003), Noll (2001a:111–147), Greeley (1977) and Kane (1955). Outside of Northern Ireland, Canada has been the most fertile ground for the Orange Order. See Houston and Smith (1980) and See (1993).

16. There are some conceptual difficulties in quantifying the number of evangelicals in any given society. Noll evaluates the difficulties survey

researchers face when attempting to distinguish the evangelicals within any given society (2001a:29–32). He isolates three strategies: One is to ascertain how many people tell survey researchers that they embrace traditional evangelical convictions concerning the Bible, the new birth, and related matters. This method is the technique regularly used by the Gallup Organization in asking people if they have been born again. In somewhat more detail, it has also been used for several recent surveys conducted by the Angus Reid Group of Toronto. A second method is to count the people adhering to the churches and denominations most strongly linked to the historical evangelical and revival movements....A third method is to figure out how many people use the term "evangelical" to describe their own religious beliefs and practices. (Noll 2001a:29)

17. Noll says it is "tempting" to equate "true believers" with "evangelicals." However, he concludes that "because the term can be used in so many different ways, it is probably more accurate to call those falling into this Angus Reid category something like Protestant 'True Believers' Who Come Close to Traditional Evangelical Definitions" (Noll 2001a:40). That said, other methods to measure the percentage of evangelicals do not produce percentages that differ radically from the "true believers" (Noll 2001a: 32–36). Accordingly, this research accepts the "true believers" percentages as sufficiently accurate for comparison of evangelicalism in the USA and Canada. The calculations of percentages of evangelicals in Northern Ireland are based on mixtures of the methods that Noll discusses.

18. The UK was included, with 5 percent of Protestants ranked as "true believers" (Noll 2001a:41).

19. Other studies of Canadian evangelicalism put the figure slightly higher. Rawlyk (1997) reports 15 percent of Canadians are evangelical. He says that about one-third of those are Catholic evangelicals. Simpson and MacLeod estimate that 9–22 percent of the population is evangelical (including Catholics and Protestants) (1985:226). Stiller calculates that 10 percent of the population is Protestant evangelicals (1991a:28).

20. Leaders of revival in the British Isles, such as Wesley and Whitefield, traveled to the American colonies where revival soon spread and was called the "Great Awakening." The beginning of the Americans' first Great Awakening is usually said to have occurred in Northampton, MA, in 1734, under the influence of Jonathan Edwards. See Ahlstrom (1972), "Jonathan Edwards and the Renewal of New England Theology," in *A Religious History of the American People.*

21. There is a case to be made that Northern evangelicals' opposition to slavery is an example of religion playing a prophetic role. This

demonstrates that even if religion has a privileged position in society, it is not inevitable that it play a priestly role.

22. William Jennings Bryan ran for president on the Democratic ticket and was leader of the Democratic Party until 1912. He served as Woodrow Wilson's secretary of state until 1915. See Marsden 1980: 132–135.

23. Seymour Martin Lipset's comparison of the impact of religion in the USA and Canada can nearly be reduced to the observation that the USA has been dominated by the values of Puritanism-evangelicalism while Canada has not (1990:88).

24. Catholics in Quebec received full British "civil rights" via the Quebec Act of 1774, fifty-five years before Catholics in Britain did (Noll 2001b:124).

25. New Brunswick was seemingly a model of peaceful religious pluralism by the mid-eighteenth century, as toleration was extended to Protestants, Catholics, and Jews—up until the expulsion of (Catholic) Acadians in 1755.

26. For instance, before the War of 1812 largely evangelical denominations such as the Baptists and Methodists had been similar to those in America. But the war nearly ended the flow of American evangelists across the border, and Canadian evangelicalism developed along more "decorous" old world lines (Noll 2001b: 267).

27. The values of slow change, consensus, and compromise also were embodied in the effort to establish a nationwide united church. Denominational boundaries had been weakening for a number of years after confederation. The tendency of Canadian evangelicals to act as mediators made them significant contributors to the United Church movement. The United Church was ultimately formed out of about "4,800 Methodist congregations, 3,700 Presbyterian congregations, 166 Congregationalist churches, and a number of union churches already existing in the west" (Noll 2001b:282).

28. These changes included an increasing role for the government in civil society, increased levels of education, and increased religious and ethnic pluralism.

29. Several volumes trace the fortunes of the Christian Right in American elections: Green, Rozell, and Wilcox 2003, 2000; Rozell and Wilcox 1997, 1995. These volumes also include case studies of the Christian Right in various states, allowing for analysis of some of the regional features and characteristics of American evangelical politics.

30. For more on the thought and practice of the evangelical "left," see Wallis (2005, 1995), Campolo (2004), and Sider (2005).

31. For instance, Reimer (2003) notes that 93.8 percent of his Canadian "core evangelical" sample attend church weekly or more; whereas 90.9 percent of his American "core evangelical" sample do so. Canadian and American evangelicals perform five other public ritual practices in largely the same percentages (Reimer 2003:105).

32. The NDP is now the most "secular" of the parties, however. See Guth and Fraser (2001).
33. The Centre for the Renewal of Public Policy was founded in 1994. It is now called the Centre for Cultural Renewal.
34. Stackhouse contrasts the EFC to the NAE in the USA. Unlike the NAE, it has never had to distance itself from fundamentalism—largely because fundamentalism has not been as prominent in Canada as in the USA. Haiven (1984), on the other hand, attempts to draw some links between the American and Canadian "born again movement." The scholarly consensus seems to be that most Canadian evangelicals do not fit into Haiven's mould.
35. Reimer finds that it is likely that the majority of evangelicals in Canada, as in the USA, are relatively disengaged from politics or even withdrawn altogether from it.
36. On evangelicalism in Ireland, see Dunlop (2004). On the role of evangelicals in the Republic of Ireland, see the *Irish Times* "Rite and Reason" column by Evangelical Alliance's Ireland director Sean Mullan (May 24, 2004). See also the Web site of EA-Ireland, http://www.evangelical.ie.
37. It is important to point out that being "anti-Agreement" can mean a number of things. For instance, it may mean that one opposes the way that the agreement has been implemented, or it may mean that one opposes the principles underlying the agreement. This distinction is not always clear-cut, within political parties or even within individuals. However, the evidence of adaptation that has taken place within both traditional evangelicalism and the DUP seems to indicate that broader, "anti-Agreement unionism" cannot be understood to simply mean that unionists oppose the power-sharing principles of the agreement (Farrington 2006).

Chapter 4   Evangelical Congregations and Identity Change

1. Ammerman et al. (1998) developed their congregational frames by drawing on the more general work of Bolman and Deal (1991), which dealt with organizations. Ammerman et al.'s frames also have similarities with the framing processes utilized in the study of new social movements (see especially McAdam, McCarthy, and Zald 1996). However, Ammerman et al. insist that their frames are specific to the study of congregations.
2. See appendix.
3. Some theologically traditional Presbyterian churches sing only Psalms, and/or use only the King James Version of the Bible.
4. The pastor said, "I think the vast majority [of members of the congregation] were no...to be perfectly honest over the time since the

Belfast Agreement was signed . . . I have met one person who admitted voting yes . . . and admitted that he was wrong" (Interview March 18, 2003).

5. Of course, people's accounts of their own behavior should not be accepted uncritically and treated as unproblematic. See Tilly (2002) and Runciman (1999).

6. The rector defined "baggage" as:

> The baggage would be to come to church wearing a suit. Some of it assumes you would be automatically unionist in outlook. You would have negative attitudes. [Such as] evangelicals don't drink, they don't smoke; they don't go to dances or anything like that. Which is a sort of identity that defines itself by what it doesn't do rather than by what it is. (Interview September 30, 2002).

7. Robert's reflections on mathematics may require further explanation. In his words:

> I like mathematics. There's a whole lot of things in mathematics that I don't really understand. I think I don't actually understand infinity, which is a very basic mathematic [concept]. I know why it has to be there and all of that, but I don't quite fully understand it. And I think [that relates to] the realization that not to understand bits of religion is alright. (Interview March 14, 2003)

8. Their positive attitude toward ecumenism was striking when compared to that of the Free Presbyterians and some of the rural Presbyterians. Nearly all of the Free Presbyterians identified ecumenism as one of the most dangerous developments in Northern Ireland.

9. Peace agents are an initiative supported by the PCI at a denominational level. Not every PCI congregation has a peace agent, however.

10. See http://www.onesmallstepcampaign.org.

11. Ikon founder Pete Rollins has written *How (Not) to Speak of God*, a book that details the philosophy behind ikon and provides examples of services at the Menagerie Bar. See also http://www.ikon.org.uk.

12. Charismatic (or Pentecostal) Christianity is often considered a subset of evangelicalism, in particular because charismatic Christians emphasize the importance of a personal relationship with Jesus and hold the Bible in high regard. However, the charismatic movement has distinguishing characteristics that set it apart from evangelicalism, for instance, its emphasis on the mystical, being "Spirit-filled," or speaking in tongues. With an estimated half-billion adherents worldwide (and still growing), charismatic Christianity is the fastest-growing expression of Christianity ever and it may be more accurate to classify evangelicalism as a subset of charismatic Christianity (see

Poloma 2003; Albrecht 1999; Land 1993). In Northern Ireland, charismatic Christianity is usually associated with the so-called nondenominational or "new" churches. There is also a small charismatic Catholic movement in Ireland.

13. Estimates varied about how many Zero28 and ikon activists attend church. The highest estimate of nonchurchgoers involved in Zero28 and ikon was 50 percent.

Chapter 5   Evangelicals and the Reframing of Political Projects

1. However, see Ganiel (2006b).
2. See appendix.
3. For more on the IOO, see Gray (1985), Morgan (1991), Patterson (1980), and Ganiel (2003).
4. This data was provided by George Dawson, former grand master of the IOO. There are about 75,000 members of the Orange Order in Northern Ireland. See also http://www.iloi.org.
5. Not to be confused with his son, also called Norman Porter, the author of *Rethinking Unionism* (1996) and *The Elusive Quest: Reconciliation in Northern Ireland* (2003). The younger Norman Porter has explicitly rejected "traditional" interpretations of evangelicalism (see his contribution in Thomson 1996).
6. The EPS was expelled from the UK-wide United Protestant Council in 2005. The controversy arose when a majority of United Protestant Council members in England voted to reject the application of the Royal Black Institution. EPS supported the Royal Black's application. For a fuller account, see "Sad Decline of United Protestant Council" in the April–June 2005 edition of the *Ulster Bulwark*.
7. The organization received about £4,000 in 2003 and £5,000–6,000 in 2002. These figures were supplied by EPS secretary Wallace Thompson.
8. Dawson died in May 2007, shortly after being reelected to the restored, power-sharing Assembly.
9. See Fuller (1996) and Howard (1986).
10. Figures for the UK are 30,000 individuals, 3,500 churches, and 700 organizations (Interview January 26, 2004).
11. ECONI received support from the following grant-making bodies and trusts: CRC (Core Funding), European Programme for Peace and Reconciliation (Peace 2), Community Bridges Programme (IFI), CRC (Grants), Parades Commission, Irish Government—Department of Foreign Affairs, Atlantic Philanthropies, Hope Trust, MAXCO Trust, Downshire Trust, Spring Harvest Trust, Demesne Charitable Trust, Ardbarron Trust, Greenview Trust, Lloyds TSB, Dorema Charitable Trust, and the Ireland Funds (ECONI Annual Report 2003–2004).

12. ECONI's name change and its attempt to set itself within an all-Ireland framework contrasts to that of EA, which operates within a UK framework. However, the launch of an EA in the Republic of Ireland in 2004 may strengthen links between evangelicals in the north and the south. For instance, representatives from EA-Northern Ireland were active in the foundation and launch of EA-Ireland.

13. Issue 36 (2004) was dedicated to "church and change" and included an interview with Noel Fallows on multicultural church life. ECONI received so much feedback on multiculturalism that it decided to dedicate Issue 37 (2004) to "racism and religious liberty." Back issues of *Lion and Lamb* are available at http://www.econi.org.

Chapter 6   Conclusions

1. See Bruce (2001b) for an analysis of the impact of evangelicalism on loyalist paramilitaries. However, Brewer argues that

    [o]ne factor Bruce overlooks...is the way in which conservative evangelical religion provides a moral critique of Catholicism that easily spills over into anti-Catholic violence amongst the rabid and irrational. The relative absence of a religious background amongst convicted Loyalist terrorists hardly does justification to the backcloth effect that anti-Catholic religious rhetoric has on the immature and gullible. (2003a:94–95)

2. This, too, has parallels with evangelicals in the Republican Party in the USA. The Republican Party must walk a fine line between keeping its evangelical constituency satisfied that it is advancing moral issues, whilst simultaneously appealing to its secular constituency. As such, evangelical "influence" in the Republican Party is always qualified and evangelicals never seem to be able to get quite what they want. Hence the disappointment of some evangelicals at the performance of Ronald Reagan, who they concluded ignored their moral concerns once they helped him get elected.

3. Dixon identifies these constraints as the need to

    sustain bipartisanship and manage domestic opinion; manage the sometimes resistant machinery of the state; contain intra-party pressures; "balance" the claims of both nationalists and unionists; maintain security and stability; appease international opinion; and secure the support and co-operation of the Republic of Ireland in dealing with violence, reducing northern nationalist alienation and promoting a stable constitutional settlement. (2001a:15, at http://www.nipolitics. com, accessed October 10, 2004)

Bibliography

Acheson, Nicholas, Brian Harvey, James Kearney, and Arthur Williamson. 2004. *Two Paths, One Purpose: Voluntary Action in Ireland—North and South.* Dublin: Institute of Public Administration.

Ahlstrom, Sydney. 1972. *A Religious History of the American People.* New Haven: Yale University Press.

Akenson, Donald. 1992. *God's People: Covenant and Land in South Africa, Israel and Ulster.* Ithaca: Cornell University Press.

Albrecht, Daniel. 1999. *Rites in the Spirit: A Ritual Approach to Pentecostal/ Charismatic Spirituality.* Sheffield: Sheffield Academic Press.

Almond, Gabriel, Scott Appleby, and Emmanuel Sivan. 2003. *Strong Religion: The Rise of Fundamentalisms around the World.* Chicago: University of Chicago Press.

Ammerman, Nancy. 2001. *Congregation and Community.* New Brunswick: Rutgers University Press.

———. 1997. "Organised Religion in a Voluntaristic Society." *Sociology of Religion* 58(3):203–215.

———. 1994. "Telling Congregational Stories." *Review of Religious Research* 35(4):189–201.

Ammerman, Nancy, Jackson Carroll, Carl Dudley, and William McKinney, eds. 1998. *Studying Congregations: A New Handbook.* Nashville: Abingdon Press.

Appleby, Scott. 2000. *The Ambivalence of the Sacred: Religion, Violence and Reconciliation.* New York: Rowman & Littlefield.

Arweck, Elisabeth, and Martin Stringer, eds. 2002. *Theorizing Faith: The Insider/Outsider Problem in the Study of Ritual.* Birmingham: University of Birmingham Press.

Aughey, Arthur. 2002. "The Art and Effect of Political Lying in Northern Ireland." *Irish Political Studies* 17(2):1–16.

———. 2001. "Learning from the Leopard." In Rick Wilford, ed. *Aspects of the Belfast Agreement.* Oxford: Oxford University Press.

Bacon, Derek. 2003. *Communities, Churches and Social Capital in Northern Ireland.* Coleraine: Centre for Voluntary Action Studies, University of Ulster.

Bacon, Derek. 1998. "Splendid and Disappointing: Churches, Voluntary Action and Social Capital in Northern Ireland." Coleraine: Centre for Voluntary Action Studies, University of Ulster.

Baudrillard, Jean. 1988. *America*. New York: Verso.

Bebbington, David. 1997. "A View from Britain." In George Rawlyk, ed. *Aspects of the Canadian Evangelical Experience*. Montreal, QC: McGill-Queen's University Press.

———. 1989. *Evangelicalism in Modern Britain: A History from the 1730s to the 1980s*. London: Unwin Hyman.

Berger, Peter. 1967. *The Sacred Canopy*. New York: Anchor Books.

Bibby, Reginald. 1993. *Unknown Gods: The Ongoing Story of Religion in Canada*. Toronto: Stoddart Publishing.

———. 1990. *Fragmented Gods: The Poverty and Potential of Religion in Canada*. Toronto: Stoddart Publishing.

"Biblical Faith for a Changing World." 2005. Belfast: Centre for Contemporary Christianity in Ireland.

Bloomfield, David. 1996. *Peacemaking Strategies in Northern Ireland: Building Complementarity in Conflict Management Theory*. Basingstoke: Palgrave Macmillan.

Boal, Frederick, Margaret Keane, and David Livingstone. 1997. *Them and Us: Attitudinal Variation among Churchgoers in Belfast*. Belfast: Institute of Irish Studies, Queen's University Belfast.

Bolman, Lee, and Terrence Deal. 1991. *Reframing Organizations: Artistry, Choice, and Leadership*. San Francisco: Jossey-Bass Inc.

Brewer, John. 2003a. *C. Wright Mills and the Ending of Violence*. Basingstoke: Palgrave Macmillan.

———. 2003b. "Northern Ireland: Peacemaking among Protestants and Catholics." In Mary Ann Cejka and Tomas Bamat, eds. *Artisans of Peace: Grassroots Peacemaking among Christian Communities*. Maryknoll, NY: Orbis Books.

———. 2002. "Are There Any Christians in Northern Ireland?" In Katrina Lloyd, Paula Devine, Gillian Robinson and Deidre Heenan, eds. *Social Attitudes in Northern Ireland: The 8th Report*. London: Pluto Press.

———. 2000. *Ethnography*. Buckingham: Open University Press.

Brewer, John, Ken Bishop, and Gareth Higgins. 2001. *Peacemaking among Protestants and Catholics*. Belfast: Centre for the Social Study of Religion, Queen's University Belfast.

Brewer, John, and Gareth Higgins). 1998. *The Mote and the Beam: Anti-Catholicism in Northern Ireland, 1600–1998*. Basingstoke: Macmillan Press.

Brubaker, Rogers. 2002. "Ethnicity without Groups." *Archives Européennes de Sociologie* 42(2):163–189.

Bruce, Steve. 2001a. "Christianity in Britain, R. I. P." *Sociology of Religion* 62(2):191–203.

———. 2001b. "Fundamentalism and Political Violence: The Case of Paisley and Ulster Evangelicals." *Religion* 31(4):387–405.

————. 1995. *Religion in Modern Britain*. Oxford: Oxford University Press.

————. 1988. *The Rise and Fall of the New Christian Right*. Oxford: Oxford University Press.

————. 1986. *God Save Ulster! The Religion and Politics of Paisleyism*. Oxford: Clarendon Press.

Buckland, Patrick. 1981. *A Short History of Northern Ireland*. New York: Holmes and Meier.

Buell, Emmett, and Lee Sigelman. 1987. "A Second Look at Popular Support for the Moral Majority." *Social Science Quarterly* 68:167–169.

————. 1985. "An Army That Meets Every Sunday? Popular Support for the Moral Majority." *Social Science Quarterly* 66:426–434. "Building on Progress: Priorities and Plans for 2003–2006." 2003. Belfast: Office of the First Minister and Deputy First Minister.

Campolo, Tony. 2004. *Speaking My Mind: The Radical Evangelical Prophet Tackles the Tough Issues Christians Are Afraid to Face*. Nashville: W Publishing Group.

"Can Do Better: Educational Disadvantage in the Context of Lifelong Learning." 2002. Belfast: Civic Forum.

Cannell, Linda. 1996. "Selective Amnesia: A Church in Transition." In Kevin Quast and John Vissers, eds. *Studies in Canadian Evangelical Renewal: Essays in Honour of Ian S. Rennie*. Toronto, ON: Faith Today Publications.

Carpenter, Joel A. 1984. "From Fundamentalism to the New Evangelical Coalition." In George Marsden, ed. *Evangelicalism and Modern America*. Grand Rapids, MI: Wm. B. Eerdmans Publishing.

Carter, Craig. 2001. *The Politics of the Cross: The Theology and Social Ethics of John Howard Yoder*. Grand Rapids, MI: Brazos Press.

Carter, Stephen. 1993. *The Culture of Disbelief: How American Law and Politics Trivialize Religious Devotion*. New York: Basic Books.

Casanova, Jose. 1994. *Public Religions in the Modern World*. Chicago: University of Chicago Press.

Choquette, Robert. 2004. *Canada's Religions: An Historical Introduction*. Ottawa: University of Ottawa Press.

Clifford, James, and George E. Marcus, eds. 1986. *Writing Culture: The Poetics and Politics of Ethnography*. Berkeley: University of California Press.

Cochrane, Feargal. 2004. "Be Careful What You Wish For. Social Capital, Civil Society and Associational Democracy: Some Lessons from Northern Ireland." Paper presented at the conference "Interpreting Ongoing Crises in the Northern Ireland Peace Process: Civil Society Dimensions." Queen's University Belfast, September 30, 2004.

————. 2001. "Unsung Heroes? The Role of Peace and Conflict Resolution Organisations in the Northern Ireland Conflict." In John McGarry, ed. *Northern Ireland and the Divided World*. Oxford: Oxford University Press.

Cochrane, Feargal, and Seamus Dunn. 2002. *People Power? The Role of the Voluntary and Community Sector in the Northern Ireland Conflict.* Cork: Cork University Press.

Cohen, Joshua. 1998. "Democracy and Liberty." In Jon Elster, ed. *Deliberative Democracy.* Cambridge: Cambridge University Press.

"Community Relations: A Brief Guide." n.d. Belfast: Community Relations Council.

"Compact: Between Government and the Voluntary and Community Sector in Northern Ireland—Building Real Partnerships." 1998. Belfast: Department of Health and Social Services.

Cooke, Dennis. 1996. *Persecuting Zeal: A Portrait of Ian Paisley.* Dingle: Brandon Book Publishers.

Cunningham, Michael. 1991. *British Government Policy in Northern Ireland 1969–89: Its Nature and Execution.* Manchester: Manchester University Press.

Davie, Grace. 1994. *Religion in Britain since 1945: Believing Without Belonging.* Oxford: Blackwell Publishers.

Dekar, Paul R. 1982. "On the Soul of Nations: Religion and Nationalism in Canada and the United States." In George Rawlyk, ed. *The Canadian Protestant Experience, 1760–1990.* Burlington, ON: Welch Publishing.

Demerath, N.J., III. 2001. *Crossing the Gods: World Religions and Worldly Politics.* New Brunswick, NJ: Rutgers University Press.

Denzin, Norman K., and Yvonna S. Lincoln, eds. 1998a. *Collecting and Interpreting Qualitative Materials.* Thousand Oaks, CA: Sage Publications.

———, eds. 1998b. *The Landscape of Qualitative Research: Theories and Issues.* Thousand Oaks, CA: Sage Publications.

———, eds. 1998c. *Strategies of Qualitative Inquiry.* Thousand Oaks, CA: Sage Publications.

Dionne Jr., E.J., and John Diiulio, Jr., eds. 2000. *What's God Got to Do with the American Experiment?* Washington, DC: Brookings Institution Press.

Directory of Cross Community Church Groups and Projects in Northern Ireland. 1999. Belfast: Community Relations Council.

Dixon, Paul. 2005. "Why the Good Friday Agreement in Northern Ireland Is Not Consociational." *Political Quarterly* 76(3):357–367.

———. 2001a. "British Policy towards Northern Ireland 1969–2000: Continuity, Tactical Adjustment and Consistent 'Inconsistencies.'" *British Journal of Politics and International Relations.* 3(3):340–368.

———. 2001b. *Northern Ireland: The Politics of War and Peace.* Basingstoke: Palgrave Macmillan.

———. 1997. "Paths to Peace in Northern Ireland (I) Civil Society and Consociational Approaches." *Democratization* 4(2):1–27.

Dudley Edwards, Ruth. 1999. *The Faithful Tribe: An Intimate Portrait of the Loyal Institutions.* London: HarperCollins.

Dunlop, Robert, ed. 2004. *Evangelicals in Ireland: An Introduction.* Blackrock, Co. Dublin: Columba Press.

Durkheim, Emile. 1968 [1915]. *The Elementary Forms of the Religious Life.* Joseph Ward Swain, trans. London: George Allen & Unwin Ltd.

"ECONI Annual Report, 2003–2004." 2004. Belfast: Evangelical Contribution on Northern Ireland.

"Education Disadvantage: A Civic Discussion." 2002. Belfast: Civic Forum.

Edwards, Michael. 2004. *Civil Society.* Cambridge: Polity Press.

Eyerman, Ron, and Andrew Jamison. 1991. *Social Movements: A Cognitive Approach.* Cambridge: Polity Press.

Fahey, Tony, Bernadette Hayes, and Richard Sinnott. 2004. *Two Traditions, One Culture? A Study of Attitudes and Values in the Republic of Ireland and Northern Ireland.* Dublin: Institute of Public Administration.

Fairclough, Norman. 2003. *Analysing Discourse: Textual Analysis for Social Research.* London: Routledge.

Farrington, Christopher. 2006. *Ulster Unionism and the Peace Process in Northern Ireland.* Basingstoke: Palgrave Macmillan.

———. 2004a. "The Democratic Unionist Party and the Northern Ireland Peace Process." Paper presented to the Political Studies Association Annual Conference, Lincoln, UK.

———. 2004b. "Models of Civil Society and Their Implications for the Northern Ireland Peace Process." IBIS Working Paper No. 43. Dublin: Institute for British-Irish Studies.

———. 2001. "Ulster Unionist Political Divisions in the Late Twentieth Century." *Irish Political Studies* 16(1):49–71.

Feuerbach, Ludwig. 1967. *Lectures on the Essence of Religion.* Ralph Manheim, trans. New York: Harper and Row.

———. 1957. *The Essence of Christianity.* George Eliot, trans. New York: Harper Torchbooks.

Fitzduff, Mari. 2002. *Beyond Violence: Conflict Resolution Process in Northern Ireland.* New York: United Nations University Press.

"For God and His Glory Alone: A Contribution Relating Some Biblical Principles to the Situation in Northern Ireland." 1988. Belfast: Evangelical Contribution on Northern Ireland.

Foucault, Michel. 1979. *Discipline and Punishment: The Birth of the Prison.* New York: Vintage.

———. 1978 *The History of Sexuality, Vol. 1.* New York: Pantheon.

Fuller, Harold. 1996. *People of the Mandate: The Story of the World Evangelical Fellowship.* Grand Rapids, MI: Baker Publishing Group.

Fulton, John. 1991. *The Tragedy of Belief: Division, Politics and Religion in Ireland.* Oxford: Clarendon Press.

Ganiel, Gladys. 2007. "Preaching to the Choir? An Analysis of DUP Discourses about the Northern Ireland Peace Process." *Irish Political Studies* 22(3):303–320.

———. 2006a. "Emerging from the Evangelical Subculture in Northern Ireland: A Case Study of the Zero28 and ikon Community." *International Journal for the Study of the Christian Church* 6(1):38–48.

Ganiel, Gladys. 2006b. "Ulster Says Maybe: The Restructuring of Evangelical Politics in Northern Ireland." *Irish Political Studies* 21(2):137–155.

Ganiel, Gladys. 2003. "The Politics of Religious Dissent in Northern Ireland." IBIS Working Paper No. 32. Dublin: Institute for British-Irish Studies.

———. 2002. "Conserving or Changing? The Theology and Politics of Northern Ireland Fundamentalist and Evangelical Protestants after the Good Friday Agreement." IBIS Working Paper No. 20. Dublin: Institute for British-Irish Studies.

Ganiel, Gladys, and Claire Mitchell. 2006. "Turning the Categories Inside-Out: Complex Identifications and Multiple Interactions in Religious Ethnography." *Sociology of Religion* 67(1):3–21.

Ganiel, Gladys, and Paul Dixon. 2008. "Religion in Northern Ireland: Rethinking Fundamentalism and the Possibilities for Conflict Transformation." *Journal of Peace Research* 45(3).

Gasaway, Brant. 2004. "Indivisible in the Work of the Kingdom: Progressive Evangelicals' Commitment to Evangelism and Social Action." Paper presented at the annual meeting of the Religious Research Association, Kansas City, MO, October 22–24, 2004.

Gavreau, Michael. 1991. *The Evangelical Century: College and Creed in English Canada from the Great Revival to the Great Depression.* Montreal, QC: McGill-Queen's University Press.

Geertz, Clifford. 1973. *The Interpretation of Cultures.* New York: Basic Books.

Gibbs, Eddie, and Ryan Bolger. 2005. *Emerging Churches: Creating Christian Community in Postmodern Cultures.* Grand Rapids, MI: Baker Academic.

Giddens, Anthony. 1998. *The Third Way.* Cambridge: Polity Press.

Gopin, Marc. 2000. *Between Eden and Armageddon: The Future of World Religions, Violence, and Peacemaking.* Oxford: Oxford University Press.

Gordon, D. 1987. "Getting Close by Staying Distant: Fieldwork with Proselytising Groups." *Qualitative Sociology* 10(3):267–287.

Gormally, Brian. 2001. "Conversion from War to Peace: Reintegration of Ex-Prisoners in Northern Ireland." BICC Paper No. 18, Bonn International Centre for Conversion, available at <http://www.bicc.de/publications/papers/paper18/content.php>.

Gray, John. 1985. *City in Revolt: James Larkin and the Belfast Dock Strike of 1907.* Belfast: Blackstaff Press.

Greeley, Andrew. 1977. *An Ugly Little Secret: Anti-Catholicism in North America.* Kansas City, MO: Sheed Andrews and McMeel.

Green, John, Corwin Smidt, James Guth, and Lyman Kellstedt. 2005. "The American Religious Landscape and the 2004 Presidential Vote: Increased Polarization." Report available at <http://pewforum.org/publications/surveys/postelection.pdf>.

Green, John, Mark Rozell, and Clyde Wilcox. 2003. *The Christian Right in American Politics: Marching to the Millennium.* Washington, DC: Georgetown University Press.

————. 2000. *Prayers in the Precincts: The Christian Right in the 1998 Elections.* Washington, DC: Georgetown University Press.

Greenawalt, Kent. 1996. *Private Consciences and Public Reasons.* Oxford: Oxford University Press.

Grenville, Andrew S. 1997. "The Awakened and the Spirit-Moved: The Religious Experiences of Canadian Evangelicals in the 1990s." In George Rawlyk, ed. *Aspects of the Canadian Evangelical Experience.* Montreal, QC: McGill-Queen's University Press.

Guelke, Adrian. 2004. "The Lure of the Miracle? The South African Connection and the Northern Ireland Peace Process." Paper presented at the conference "Interpreting Ongoing Crises in the Northern Ireland Peace Process." Queen's University Belfast, June 11, 2004.

Guest, Matthew. 2002. "'Alternative' Worship: Challenging the Boundaries of the Christian Faith." In Elisabeth Arweck and Martin Stringer, eds. *Theorizing Faith: The Insider/Outsider Problem in the Study of Ritual.* Birmingham: University of Birmingham Press.

Guth, James L. 1996. "The Politics of the Christian Right." In John Green, James L. Guth, Corwin E. Smidt, Lyman A. Kellstedt, and John Clifford Green, eds. *Religion and the Culture Wars: Dispatches from the Front.* Lanham, MD: Rowman & Littlefield.

Guth, James L., and Cleveland Fraser. 2001. "Religion and Partisanship in Canada." *Journal for the Scientific Study of Religion* 40(1):51–64.

Guth, James L., John Green, Lyman Kellstedt, and Corwin Smidt. 1996. "Onward Christian Soldiers: Religious Activist Groups in American Politics." In John Green, James L. Guth, Corwin E. Smidt, Lyman A. Kellstedt, and John Clifford Green, eds. *Religion and the Culture Wars: Dispatches from the Front.* Lanham, MD: Rowman & Littlefield.

Haddick-Flynn, Kevin. 1999. *Orangeism: The Making of a Tradition.* Dublin: Wolfhound Press.

Haiven, Judy. 1984. *Faith, Hope, No Charity.* Vancouver, BC: New Star Books.

Handy, Robert T. 1982. "Protestant Patterns in Canada and the United States: Similarities and Differences." In Joseph D. Ban and Paul R. Dekar, eds. *In the Great Tradition: In Honour of Winthrop S. Hudson, Essays on Pluralism, Voluntarism, and Revivalism.* Valley Forge, PA: Judson Press.

Hann, Chris. 1996. "Introduction: Political Society and Civil Anthropology." In Chris Hann and Elizabeth Dunn, eds. *Civil Society: Challenging Western Models.* London: Routledge.

Hauerwas, Stanley. 1995. *In Good Company: The Church as Polis.* Notre Dame, IN: University of Notre Dame Press.

————. 1983. *The Peaceable Kingdom: A Primer in Christian Ethics.* Notre Dame, IN: University of Notre Dame Press.

Hauerwas, Stanley, and William H. Willimon. 1989. *Resident Aliens: Life in the Christian Colony: A Provocative Christian Assessment of Culture and Ministry for People Who Know That Something Is Wrong.* Nashville, TN: Abingdon Press.

Hauerwas, Stanley, and William H. Willimon. 1995. "Why Resident Aliens Struck a Chord." In Stanley Hauerwas, ed. *In Good Company: The Church As Polis.* Notre Dame, IN: University of Notre Dame Press.

Hayes, Bernadette, and Ian McAllister. 1999. "Ethnonationalism, Public Opinion and the Good Friday Agreement." In Joseph Ruane and Jennifer Todd, eds. *After the Good Friday Agreement.* Dublin: University College Dublin Press.

Hefner, Robert W. 2001. *The Politics of Multiculturalism: Pluralism and Citizenship in Malaysia, Singapore and Indonesia.* Honolulu: University of Hawaii Press.

———. 2000. *Civil Islam: Muslims and Democratization in Indonesia.* Princeton, NJ: Princeton University Press.

———, ed. 1998. *Democratic Civility: The History and Cross-cultural Possibility of a Modern Political Ideal.* New Brunswick, NJ: Transaction Publishers.

Heimert, Alan. 1966. *Religion and the American Mind: From the Great Awakening to Revolution.* Cambridge, MA: Harvard University Press.

Hempton, David, and Myrtle Hill. 1992 *Evangelical Protestantism in Ulster Society 1740–1890.* London: Routledge.

Herbert, David. 2003. *Religion and Civil Society: Rethinking Religion in the Contemporary World.* Aldershot: Ashgate Publishing.

Heskin, Ken. 1980. *Northern Ireland: A Psychological Analysis.* Dublin: Gill and Macmillan.

Hessel, Dieter, ed. 1993. *The Church's Public Role.* Grand Rapids, MI: Wm. B. Eerdmans Publishing.

Houston, Cecil, and William Smyth. 1980. *The Sash Canada Wore: A Historical Geography of the Orange Order in Canada.* Toronto, ON: University of Toronto Press.

Howard, David. 1986. *The Dream That Would Not Die: The Birth and Growth of the World Evangelical Fellowship, 1846–1986.* Exeter: Paternoster Press.

Hunter, James Davison. 1991. *Culture Wars: The Struggle to Define America.* New York: Basic Books.

———. 1987. *Evangelicalism: The Coming Generation.* Chicago: University of Chicago Press.

"Investing Together: Report of the Taskforce on Resourcing the Voluntary and Community Sector." 2004. Belfast: Department of Social Development.

James, Allison, Jenny Hockey, and Andrew Dawson, eds. 1997. *After Writing Culture: Epistemology and Praxis in Contemporary Anthropology.* New York: Routledge.

Jelen, Ted. 2002. *Sacred Markets, Sacred Canopies: Essays on Religious Markets and Religious Pluralism.* New York: Rowman & Littlefield

———. 2000. *To Serve God and Mammon: Church-State Relations in American Politics.* Boulder, CO: Westview Press.

———. 1991. *The Political Mobilization of Religious Beliefs.* Westport, CT: Praeger.

Jelen, Ted, and Clyde Wilcox. 2002. *Religion and Politics in Comparative Perspective.* Cambridge: Cambridge University Press.

Johnston, Douglas, ed. 2003. *Trumping Realpolitik: Faith-Based Diplomacy.* Oxford: Oxford University Press.

Jordan, Glenn. 2001. *Not of This World: The Evangelical Protestants of Northern Ireland.* Belfast: Blackstaff Press.

Kaldor, Mary. 2003. *Global Civil Society: An Answer to War.* Cambridge: Polity Press.

Kaufmann, Eric. 2007. *The Orange Order: A Contemporary Northern Irish History.* Oxford: Oxford University Press.

Keane, John J. 1998. *Civil Society: Old Images, New Visions.* Cambridge: Polity Press.

———. 1996. *Reflections on Violence.* London: Verso.

———. 1988a. *Democracy and Civil Society: On the Predicaments of European Socialism, the Prospects for Democracy, and the Problem of Controlling Social and Political Power.* London: Verso.

———, ed. 1988b. *Civil Society and the State: New European Perspectives.* London: Verso.

———. 1955. *Catholic-Protestant Conflicts in America.* Chicago: Regenery.

Kearney, James R., and Arthur P. Williamson. 2001. "The Voluntary and Community Sector in Northern Ireland: Developments since 1995–96." In *Next Steps in Voluntary Action: An Analysis of Five Years of Development in the Voluntary Sector in England, Northern Ireland, Scotland and Wales.* London: Centre for Civil Society, London School of Economics and National Council of Voluntary Organisations.

Kellstedt, Lyman, John Green, James Guth, and Corwin Smidt. 1994. "Religious Voting Blocs in the 1992 Election: The Year of the Evangelical?" *Sociology of Religion* 55:307–326.

Kymlicka, Will. 1995. *Multicultural Citizenship: A Liberal Theory of Minority Rights,* Oxford: Clarendon.

Land, Steven. 1993. *Pentecostal Spirituality.* Sheffield: Sheffield Academic Press.

Langhammer, Mark. 2004. "State Funded Sectarianism and Pandering to Paramilitarism." Paper presented at the conference "Interpreting Ongoing Crises in the Northern Ireland Peace Process: Civil Society Dimensions." Queen's University Belfast, September 30, 2004.

Larkin, Emmet. 1976. *The Historical Dimensions of Irish Catholicism.* Dublin: Four Courts Press.

Leege, David. 1993. "Religion and Politics in Theoretical Perspective." In David Leege and Lyman Kellstedt, eds. *Rediscovering the Religious Factor in American Politics.* Armonk, NY: M.E. Sharpe.

Levinson, Sanford. 1992. "Religious Language and the Public Square." *Harvard Law Review* 105(8):2061–2079.

Liebman, Robert, and Robert Wuthnow, eds. 1983. *The New Christian Right.* New York: Aldine Publishing.

Liechty, Joseph, and Cecelia Clegg. 2001. *Moving beyond Sectarianism: Religion, Conflict and Reconciliation in Northern Ireland.* Dublin: Columba Press.

Lijphart, Arend. 1977. *Democracy in Plural Societies.* London: Yale University Press.

Lindsay, D. Michael. 2004. "Elite Networks as Social Power: New Modes of Organisation within American Evangelicalism." Paper presented at the annual meeting of the Society for the Scientific Study of Religion, Kansas City, MO. October 22–24, 2004.

Lindsay, Isobel. 2000. "The New Civic Forums." *Political Quarterly* 71(4):404–411.

Lipset, Seymour Martin. 1990. *Continental Divide.* New York: Routledge.

Little, Adrian. 2004. *Democracy and Northern Ireland: Beyond the Liberal Paradigm?* Basingstoke: Palgrave Macmillan.

Little, David, ed. 2007. *Peacemakers in Action: Profiles of Religion in Conflict Resolution.* Cambridge: Cambridge University Press.

Livingstone, Stephen, and Rachel Murray. 2005. "Evaluating the Effectiveness of National Human Rights Institutions: The Northern Ireland Human Rights Commission with Comparisons from South Africa." Nuffield Foundation Report.

Lofland, J.A., and L.H. Lofland. 1984. *Analyzing Social Settings: A Guide to Qualitative Observation and Analysis.* Belmont, CA: Wadsworth Publishing.

Lukens-Bull, Ronald. 2005. *A Peaceful Jihad: Negotiating Identity and Modernity in Muslim Java.* New York: Palgrave Macmillan.

Lyotard, Jean-François. 1979. *The Postmodern Condition.* Minneapolis, MN: University of Minneapolis Press.

Machen, J. Gresham. 1923. *Christianity and Liberalism.* Grand Rapids, MI: Wm. B. Eerdmans Publishing.

"Making a Difference 2002–2005." 2001. Belfast: The Northern Ireland Executive's First Programme for Government, Office of the First Minister and Deputy First Minister.

Marsden, George. 1980. *Fundamentalism and American Culture: The Shaping of 20th Century Evangelicalism 1870–1925.* Oxford: Oxford University Press.

Martin, David. 1978. *A General Theory of Secularization.* Oxford: Basil Blackwell.

Marty, Martin (with Jonathan Moore). 2000. *Politics, Religion and the Common Good: Advancing a Distinctly American Conversation about Religion's Role in Our Shared Life.* San Francisco: Jossey-Bass Inc.

Marty, Martin, and R. Scott Appleby, eds. 1995. *Fundamentalisms Comprehended.* Chicago: University of Chicago Press.

———, eds. 1994. *Accounting for Fundamentalisms.* Chicago: University of Chicago Press.

———, eds. 1993a. *Fundamentalisms and Society.* Chicago: University of Chicago Press.

————, eds. 1993b. *Fundamentalisms and the State*. Chicago: University of Chicago Press.

————, eds. 1992. *Fundamentalisms Observed*. Chicago: University of Chicago Press.

Marx, Karl, and Frederick Engels. 1975. *On Religion*. Moscow: Progress Publishers.

Massa, Mark. 2003. *Anti-Catholicism in America: The Last Acceptable Prejudice*. New York: Crossroad General Interest.

McAdam, Doug, John McCarthy, and Mayer Zald. 1996. *Comparative Perspectives on Social Movements: Political Opportunities, Mobilizing Structures, and Cultural Framings*. Cambridge: Cambridge University Press.

McBride, Ian. 1998. *Scripture Politics: Ulster Presbyterians and Irish Radicalism in the Late Eighteenth Century*. Oxford: Clarendon Press.

McCrudden, Christopher. 1999. "Equality and the Good Friday Agreement." In Joseph Ruane and Jennifer Todd, eds. *After the Good Friday Agreement*. Dublin: University College Dublin Press.

McEvoy, Kieran. 2001. *Paramilitary Imprisonment in Northern Ireland: Resistance, Management, and Release*. Oxford: Oxford University Press.

McGarry, John, and Brendan O'Leary. 2004. *The Northern Ireland Conflict: Consociational Engagements*. Oxford: Oxford University Press.

McKnight, Scot. 2007. "Five Streams of the Emerging Church: Key Elements of the Most Controversial and Misunderstood Movement in the Church Today." *Christianity Today* 51(2):35–39.

McLaren, Brian. 2004. *A Generous Orthodoxy*. Grand Rapids, MI: Zondervan.

Melucci, Alberto. 1988a. "Getting Involved: Identity and Mobilization in Social Movements." *International Social Movement Research* 1:329–348.

————. 1988b. "Social Movements and the Democratization of Everyday Life." In John Keane, ed. *Civil Society and the State: New European Perspectives*. London: Verso.

————. 1985. "The Symbolic Challenge of Contemporary Movements." *Social Research* 52(4):789–816.

Miller, David. 1978. *Queen's Rebels: Ulster Loyalism in Historical Perspective*. Dublin: Gill and Macmillan.

Mitchel, Patrick. 2003. *Evangelicalism and National Identity in Ulster, 1921–1998*. Oxford: Oxford University Press.

Mitchell, Claire. 2006. *Religion, Identity and Politics in Northern Ireland*. Aldershot: Ashgate Publishing.

————. 2003. "Protestant Identification and Political Change in Northern Ireland." *Ethnic and Racial Studies* 26(4):612–631.

Mitchell, Claire, and James Tilley. 2004. "The Moral Minority: Evangelical Protestants in Northern Ireland and Their Political Behaviour." *Political Studies* 52(4):585–602.

Moen, Matthew C. 1992. *The Transformation of the Christian Right*. Tuscaloosa: University of Alabama Press.

Moloney, Ed. 2002. *A Secret History of the IRA*. London: Allen Lane.

Moloney, Ed, and Andy Pollak. 1986. *Paisley*. Swords: Poolbeg Press.

Monsma, Stephen, and J. Christopher Soper. 1997. *The Challenge of Pluralism: Church and State in Five Democracies*. New York: Rowman & Littlefield.

Morgan, Austen. 1991. *Labour and Partition: The Belfast Working Class 1905–23*. London: Pluto Press.

Morrow, Duncan, Derek Birrell, John Greer, and Terry O'Keefe. 1991. *The Churches and Inter-community Relationships*. Coleraine: Centre for the Study of Conflict.

Mouffe, Chantal. 2000. *The Democratic Paradox*. London: Verso.

Mullan, Sean. 2004. "Creating a New Engagement with the Gospel of Jesus." *Irish Times*. May 24, 2004.

Munson, Henry. 2005. "Fundamentalism." In John Hinnells, ed. *Companion to the Study of Religion*. London: Routledge.

———. 2003. "Fundamentalism Ancient and Modern." *Daedalus* 132(3):31–42.

Nason-Clark, Nancy, and Mary Jo Neitz, eds. 2001. *Feminist Narratives in the Sociology of Religion*. Walnut Creek, CA: AltaMira Press.

Neitz, Mary Jo. 2004. "Gender and Culture: Challenges to the Sociology of Religion." *Sociology of Religion* 65(4):391–402.

———. 2002. "Walking between the Worlds: Permeable Boundaries, Ambiguous Identities." In James V. Spickard, J. Shawn Landres, and Meredith B. McGuire, eds. *Personal Knowledge and Beyond: Reshaping the Ethnography of Religion*. New York: New York University Press.

Neuhaus, Richard. 1986. *The Naked Public Square: Religion and Democracy in America*. Grand Rapids, MI: Wm. B. Eerdmans Publishing.

NICVA Response to "A Shared Future." Belfast: Northern Ireland Council for Voluntary Action, available at <http://www.nicva.org/policy> and <www.nicva.org/research/responses>

Niebuhr, H. Richard. 2001 [1951]. *Christ and Culture*. San Francisco: HarperCollins.

Noll, Mark. 2001a. *American Evangelical Christianity: An Introduction*. Oxford: Blackwell Publishers.

———. 2001b [1992]. *A History of Christianity in the United States and Canada*. Grand Rapids, MI: Wm. B. Eerdmans Publishing.

———. 1998. "Religion in Canada and the United States: Comparisons from an Important Survey Featuring the Place of Evangelical Christianity." *Crux* 34(4):13–25.

———. 1995. *The Scandal of the Evangelical Mind*. Grand Rapids, MI: Wm. B. Eerdmans Publishing

Noll, Mark, Nathan Hatch, and George Marsden. 1989. *The Search for Christian America*. Colorado Springs: Helmers & Howard.

"Northern Ireland Voluntary and Community Almanac: State of the Sector III." 2002. Belfast: Northern Ireland Council for Voluntary Action.

O'Brien, Conor Cruise. 1974. *States of Ireland*. London: Panther Books.

O'Callaghan, Margaret, and Catherine O'Donnell. 2006. "The Northern Ireland Government, the 'Paisleyite Movement' and Ulster Unionism in 1966." *Irish Political Studies* 21(2):203–222.

O'Leary, Brendan. 1997. "The Conservative Stewardship of Northern Ireland, 1979–1997: Sound-Bottomed Contradictions or Slow Learning." *Political Studies* 45(4):663–676.

O'Malley, Padraig. 1983. *The Uncivil Wars: Ireland Today*. Belfast: Blackstaff Press.

Osborne, Robert, and Ian Shuttleworth, eds. 2004. *Fair Employment in Northern Ireland: A Generation On*. Belfast: Blackstaff Press.

Palshaugen, Lone. 2004. "The Northern Ireland Civic Forum and the Politics of Recognition." Working Paper No. 38. Dublin: Institute for British-Irish Studies, University College Dublin.

———. 2002. "The Civic Forum and the Politics of Recognition: A New Arena for Conflict or a New Political Culture?" Unpublished Master's Thesis, University College Dublin.

Parekh, Bhikhu. 2000. *Rethinking Multiculturalism: Cultural Diversity and Political Theory*. London: Palgrave Macmillan.

"Partners for Change: Government's Strategy for Support of the Voluntary and Community Sector 2001–2004." 2001. Belfast: Office of the First Minister and Deputy First Minister.

Patterson, Henry. 1980. *Class Conflict and Sectarianism: The Protestant Working Class and the Belfast Labour Movement 1868–1920*. Belfast: Blackstaff Press.

Peshkin, Alan. 1984. "Odd Man Out: The Participant Observer in an Absolutist Setting." *Sociology of Education* 57(4):254–264.

Pollak, Andy, ed. 1993. *A Citizens' Inquiry: The Opsahl Report on Northern Ireland*. Dublin: Lilliput Press.

Poloma, Margaret. 2003. *Main Street Mystics*. New York: AltaMira Press.

Porter, David. 2005. "Ireland Is Changing and So Are We!" *Lion and Lamb* 38:4.

Porter, Fran. 2002. *Changing Women, Changing Worlds: Evangelical Women in Church, Community, and Politics*. Belfast: Blackstaff Press.

Porter, Norman. 2003. *The Elusive Quest: Reconciliation in Northern Ireland*. Belfast: Blackstaff Press.

———. 1996. *Rethinking Unionism*. Belfast: Blackstaff Press.

Power, Maria. 2007. *From Ecumenism to Community Relations: Inter-church Relationships in Northern Ireland, 1980–2005*. Dublin: Irish Academic Press.

———. 2005. "Building Communities in a Post-conflict Society: Churches and Peace-Building Initiatives in Northern Ireland since 1994." *The European Legacy* 10(1):55–68.

Putnam, Robert. 2000. *Bowling Alone: The Collapse and Revival of American Community*. New York: Simon and Schuster.

———. 1993. *Making Democracy Work: Civic Traditions in Modern Italy*. Princeton: Princeton University Press.

Rankin, Amber, and Gladys Ganiel. 2007. "Denouncing and Denying: DUP Discourses about Violence and Their Impact on the Peace Process." Paper presented at the Peace Lines Conference, Humanities Institute of Ireland, Dublin, June 1, 2007.

Rankin, Anna. 2005. "Spring Fever." *Lion and Lamb* 38:1.

Rawls, John. 1993. *Political Liberalism*. New York: Columbia University Press.

Rawlyk, George. 1997. *Aspects of the Canadian Evangelical Experience*. Montreal, QC: McGill-Queen's University Press.

———. 1990. *The Canadian Protestant Experience 1760–1990*. Burlington, ON: Welch Publishing.

"A Regional Strategy for Social Inclusion." Belfast: Civic Forum, 2002.

Reimer, Sam. 2003. *Evangelicals and the Continental Divide: The Conservative Protestant Subculture in Canada and the United States*. Montreal, QC: McGill-Queen's University Press.

Rollins, Peter. 2006. *How (Not) to Speak of God*. London: SPCK Publishing.

Ronson, Jon. 2002. *Them: Adventures with Extremists*. New York: Simon and Schuster.

Rose, Richard. 1971. *Governing Without Consensus: An Irish Perspective*. London: Faber and Faber.

Rozell, Mark, and Clyde Wilcox, eds. 1997. *God at the Grassroots: The Christian Right in the 1996 Elections*. Lanham, MD: Rowman & Littlefield.

———, eds. 1995. *God at the Grassroots: The Christian Right in the 1994 Elections*. Lanham, MD: Rowman & Littlefield.

Ruane, Joseph, and Jennifer Todd. Forthcoming in 2008. *Dynamics of Conflict and Transition*. Cambridge: Cambridge University Press.

———. 2004. "The Roots of Intense Ethnic Conflict May Not Themselves Be Ethnic: Categories, Communities, and Path Dependence." *Archives Européennes de Sociologie* 45(2):209–232.

———, eds. 1999. *After the Good Friday Agreement*. Dublin: University College Dublin Press.

———. 1996. *The Dynamics of Conflict in Northern Ireland*. Cambridge: Cambridge University Press.

Runciman, W.G. 1999. *The Social Animal*. London: Fontana Press.

"Sad Decline of United Protestant Council." 2005. *Ulster Bulwark,* April–June 2005.

Sandel, Michael. 1996. *Democracy's Discontent: America in Search of a Public Philosophy*. Cambridge, MA: Belknap Press.

Schatzman, Leonard, and Anselm Strauss. 1973. *Field Research: Strategies for a Natural Sociology*. Englewood Cliffs, NJ: Prentice Hall.

See, Scott. 1993. *Riots in New Brunswick: Orange Nativism and Social Violence in the 1840s*. Toronto, ON: University of Toronto Press.

"A Shared Future: Improving Relations in Northern Ireland." 2003. Belfast: Office of the First Minister and Deputy First Minister.

"A Shared Future: Policy and Strategic Framework for Good Relations in Northern Ireland." 2005. Belfast: Office of the First Minister and Deputy First Minister.

Shea, William. 2004. *The Lion and the Lamb: Evangelicals and Catholics in America*. Oxford: Oxford University Press.

Sider, Ronald. 2005. *The Scandal of the Evangelical Conscience*. Grand Rapids, MI: Baker Books.

Sigelman, Lee, Clyde Wilcox, and Emmett Buell. 1987. "An Unchanging Minority: Popular Support for the Moral Majority, 1980 and 1984." *Social Science Quarterly* 68:876–884.

Silverman, David. 1993. *Interpreting Qualitative Data: Methods for Analysing Talk, Text and Interaction*. London: Sage Publications.

Simpson, John H., and Henry G. MacLeod. 1985. "The Politics of Morality in Canada." In Rodney Stark, ed. *Religious Movements: Genesis, Exodus, and Numbers*. New York: Paragon House Publishers.

Smith, Christian. 1998. *American Evangelicalism: Embattled and Thriving*. Chicago: University of Chicago Press.

Smyth, Clifford. 1987. *Ian Paisley: Voice of Protestant Ulster*. Edinburgh: Scottish Academic Press.

Southern, Neil. 2005. "Ian Paisley and Evangelical Democratic Unionists: An Analysis of the Role of Evangelical Protestantism within the Democratic Unionist Party." *Irish Political Studies* 20(2):127–145.

Spickard, James V., J. Shawn Landres, and Meredith B. McGuire, eds. 2002. *Personal Knowledge and Beyond: Reshaping the Ethnography of Religion*. New York: New York University Press.

Stackhouse, John G., Jr. 1997 "Who Whom? Evangelicalism and Canadian Society." In George Rawlyk, ed. *Aspects of the Canadian Evangelical Experience*. Montreal, QC: McGill-Queen's University Press.

———. 1995. "The National Association of Evangelicals, the Evangelical Fellowship of Canada, and the Limits of Evangelical Cooperation." *Christian Scholar's Review* 25(2):157–179.

———. 1990. "The Protestant Experience in Canada since 1945." In George Rawlyk, ed. *The Canadian Protestant Experience, 1760–1990*. Burlington, ON: Welch Publishing.

Stark, Rodney. 1999. "Secularization, RIP." *Sociology of Religion* 60(3): 249–273.

Stiller, Brian. 1997. *From the Tower of Babel to Parliament Hill: How to Be a Christian in Canada Today*. Toronto, ON: Harper Collins.

———. 1996. "Canadian Pluralism: Friend or Foe?" In Kevin Quast and John Vissers, eds. *Studies in Canadian Evangelical Renewal: Essays in Honour of Ian S. Rennie*. Toronto, ON: Faith Today Publications.

———. 1991a. *Critical Options for Evangelicals*. Toronto, ON: Faith Today Publications.

———. 1991b. "A Personal Coda." In Robert E. Vander Vennen, ed. *Church and Canadian Culture*. Lanham, MD: University Press of America.

"Strategy for Support of the Voluntary Sector and for Community Development in Northern Ireland." 1993. Belfast: Department of Health and Social Services.

Stringer, Martin. 2002. "Introduction: Theorizing Faith." In Elisabeth Arweck and Martin Stringer, eds. *Theorizing Faith: The Insider/Outsider Problem in the Study of Ritual*, Birmingham: University of Birmingham Press.

Taylor, Rupert. 2001. "Northern Ireland: Consociation or Social Transformation?" In John McGarry, ed. *Northern Ireland and the Divided World: Post-agreement Northern Ireland in Comparative Perspective*. Oxford: Oxford University Press.

Tempest, Clive. 1997. "Myths from Eastern Europe and the Legend of the West." *Democratization* 4(2):132–144.

Thompson, Simon. 2002. "Parity of Esteem and the Politics of Recognition." *Contemporary Political Theory* 1(2):203–220.

Thomson, Alwyn. 2002. *Fields of Vision: Faith and Identity in Protestant Ireland*. Belfast: Centre for Contemporary Christianity in Ireland.

———, ed. 1999. *The Great White Tent*. Belfast: Evangelical Contribution on Northern Ireland.

———. 1998. "Evangelicalism and Fundamentalism." In Norman Richardson, ed. *A Tapestry of Beliefs: Christian Traditions in Northern Ireland*. Belfast: Blackstaff Press.

———, ed. 1996. *Faith in Ulster*. Belfast: Evangelical Contribution on Northern Ireland.

———. 1995a. *Beyond Fear, Suspicion and Hostility: Evangelical-Roman Catholic Relationships*. Belfast: Nelson and Knox.

———. 1995b. *The Fractured Family*. Belfast: Nelson and Knox.

Tilly, Charles. 2002. *Stories, Identities, and Political Change*. Oxford: Rowman & Littlefield.

Todd, Jennifer. 2005. "Social Transformation, Collective Categories and Identity Change." *Theory and Society* 35(4):429–463.

———. 1987. "Two Traditions in Unionist Political Culture." *Irish Political Studies* 2(1):1–26.

Tombs, David, and Joseph Leichty, eds. 2006. *Explorations in Reconciliation: New Directions in Theology*. Aldershot: Ashgate Publishing.

Tomlinson, Dave, and Dallas Willard. 2003. *The Post-evangelical*. Grand Rapids, MI: Zondervan.

Tonge, Jonathan. 2005. *The New Northern Irish Politics?* Basingstoke: Palgrave Macmillan.

Touraine, Alain. 1978. *The Voice and the Eye: An Analysis of Social Movements*. Cambridge: Cambridge University Press.

"Trends: Religion and Public Life—A Faith-Based Partisan Divide." 2005. Pew Forum Report available at: <http://pewforum.org/publications/reports/religion-and-politics-report.pdf>

Varshney, Ashutosh. 2003. *Ethnic Conflict and Civic Life: Hindus and Muslims in India*. New Haven: Yale University Press.

———. 2001. "Ethnic Conflict and Civil Society." *World Politics* 53: 362–398.

Wald, Kenneth D. 2003. *Religion and Politics in the United States* (4th edition). New York: Rowman & Littlefield.

Wallis, Jim. 2005. *God's Politics: Why the Right Gets It Wrong and the Left Doesn't Get It.* San Francisco: HarperSanFrancisco.

———. 1995. *The Soul of Politics: Beyond "Religious Right" and "Secular Left."* San Diego: Harvest Books.

Wallis, Roy, and Steve Bruce. 1986. *Sociological Theory, Religion and Collective Action.* Belfast: Queen's University.

Warner, R. Stephen. 1988. *New Wine in Old Wineskins: Evangelicals and Liberals in a Small-Town Church.* Berkeley: University of California Press.

Watson, Justin. 1997. *The Christian Coalition: Dreams of Restoration, Demands for Recognition.* New York: St. Martin's Press.

Weller, Robert P. 1999. *Alternate Civilities: Democracy and Culture in China and Taiwan.* Boulder, CO: Westview Press.

Wells, Ronald, and David Livingstone. 1999. *Ulster-American Religion: Episodes in the History of a Cultural Connection.* Notre Dame, IN: University of Notre Dame Press.

White, Jenny. 2002. *Islamist Mobilization in Turkey.* Seattle: University of Washington Press.

Wilcox, Clyde. 2002. "The Christian Right in the 2000 Elections." In Stephen Wayne and Clyde Wilcox, eds. *The Election of the Century and What It Tells Us about the Future of American Politics.* Armonk, NY: M.E. Sharpe.

———. 1992. *God's Warriors: The Christian Right in 20th Century America.* Baltimore: Johns Hopkins University Press.

———. 1987. "Popular Support for the Moral Majority in 1980: A Second Look." *Social Science Quarterly* 68:157–167.

Wilcox, Clyde, and Mark Rozell. 2000. "Conclusion: The Christian Right in Campaign '98." In John Green, Mark Rozell, and Clyde Wilcox, eds. *Prayers in the Precincts: The Christian Right in the 1998 Elections.* Washington, DC: Georgetown University Press.

Wilcox, Melissa. 2002. "Dancing on the Fence: Researching Lesbian, Gay, Bisexual, and Transgender Christians." In James V. Spickard, J. Shawn Landres, and Meredith B. McGuire, eds. *Personal Knowledge and Beyond: Reshaping the Ethnography of Religion.* New York: New York University Press.

Wilson, Bryan. 1979. *Contemporary Transformations of Religion.* Oxford: Oxford University Press.

Wilson, Derick, and Jerry Tyrell. 1995. "Institutions for Conciliation and Mediation." In Seamus Dunn, ed. *Facets of the Conflict in Northern Ireland.* London: Palgrave Macmillan.

Wright, Frank. 1973. "Protestant Ideology and Politics in Ulster." *European Journal of Sociology* 14(2):213–280.

Wuthnow, Robert. 1998. *After Heaven: Spirituality in America Since the 1950s*. Berkeley: University of California Press.

———. 1988. *The Restructuring of American Religion: Society and Faith since World War II*. Princeton, NJ: Princeton University Press.

Yoder, John Howard. 1994 [1972]. *The Politics of Jesus*. Grand Rapids, MI: Wm. B. Eerdmans Publishing.

Index